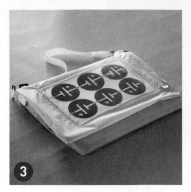

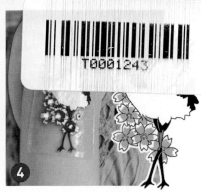

Carson Scott, Georgina Metzler, Karl Wagner, Danielle Rhodes, Katherine Connell, Jiacheng Jiang, Aurélie Sowerbutts, Ruben Agullo

Manufacturable, with a transformative design that allows the wearer to adjust to the season and even transform the garment into a tote, all while charging your mobile device using the sun's rays.

Most Creative went to **Georgina Metzler's** *Dragonfly Dress* (Figure ❷), which tells the Anishinaabe story of the Dragonfly, reminding us to be ourselves with solar-powered nLITEn LED dragonflies shining through laser-cut satin.

Solaire's *Reactive Safety Vest* impressed Hackster members with its practical flashing light patterns and haptic alerts, earning the most "likes" of any project, and with them the Most Wanted (People's Choice) award.

Karl Wagner and Danielle Rhodes' *Ultra-lightweight Solar-Powered Messenger Bag* (Figure ❸) earned Most Practical with a "well-documented prototype that is clearly manufacturable."

The first of the surprise awards went to **Katherine Connell's** *Sprite Lights* (Figure ❹), LED body art that took the unique approach of applying flexible circuits, LEDs, and solar films to "silicon skin" to create a completely new type of wearable, thus nabbing the Out of the Box prize.

Jiacheng Jiang's *Domain* features not only changing colors, but changing shapes, with app-controlled butterfly wings and dynamic length, earning the Design Dynamo award (Figure ❺).

Innovation Maverick was awarded to **Aurélie Sowerbutts'** Harajuku-inspired, fashion-forward 4-in-1 *Solar Powered LED Shirt* crafted from deadstock fabrics (Figure ❻).

ConceptCindy's whimsically-named *Fun Pants* illuminate your path and the dancefloor with LED-lined vents and a special pocket to conceal the nLITEn controller, nabbing her the Tech Trendsetter award.

Fashion meets function with **Stive H's** Palette Pioneer-winning *Solar-Powered Color-Changing Jacket*, featuring user-adjustable LEDs along the fabric's contours.

Ruben Agullo's *Muzzle Outfit* emphasizes function, with frostbite-fighting heated gloves and insoles, garnering the Future Watch prize for this utilitarian design (Figure ❼).

Semi-finalists and qualified surprise award winners will work with Art by Physicist to bring their concepts to market, culminating in a joint Kickstarter campaign, with backers' wallets determining the ultimate victors.

Subscribe to the Art by Physicist newsletter on kittyyeung.com for updates! ◗

Make: 87

46

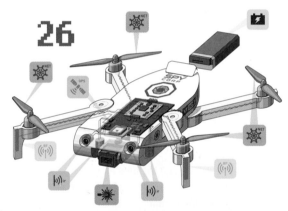

26

50

ON THE COVER:
From left to right: Arduino Uno R4 and Raspberry Pi 5. Plus, Debra Ansell's tesellated LED tote bag.

Photos: Mark Madeo and Debra Ansell

CONTENTS

SPECIAL INSERT
Make: Guide to Boards 2024
Check out specs and hands-on reviews of the latest microcontrollers and single-board computers.
80+ boards for you to compare!

Mark Madeo, Rob Nance, Matt Venn, Doug Stowe, Lee Wilkens, Chris Borge, Bob Knetzger, Annabel Allen, Charles Platt

76

88

98

110

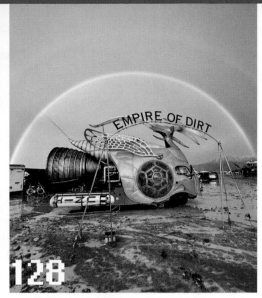
EMPIRE OF DIRT
128

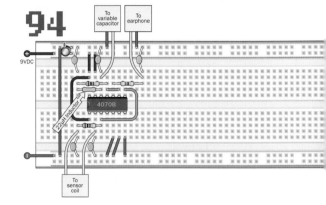
94

Make:®

> "I can do more damage on my laptop sitting in my pajamas before my first cup of Earl Grey than you can do in a year in the field."
> —Q (Skyfall)

PRESIDENT
Dale Dougherty
dale@make.co

VP, PARTNERSHIPS
Todd Sotkiewicz
todd@make.co

EDITORIAL

EDITOR-IN-CHIEF
Keith Hammond
keith@make.co

SENIOR EDITOR
Caleb Kraft
caleb@make.co

COMMUNITY EDITOR
David J. Groom
david@make.co

PRODUCTION MANAGER
Craig Couden

CONTRIBUTING EDITORS
Tim Deagan
William Gurstelle

CONTRIBUTING WRITERS
Debra Ansell, Cabe Atwell, Mohit Bhoite, Chris Borge, Rich Cameron, Christina Cole, Laurel Cummings, Greg Gilman, Diego Gonzalez, Joe Grand, Gareth Halfacree, Joan Horvath, Victoria Jaqua, Steve Johnson, Rick Kaseguma, Bob Knetzger, Nick Lambourne, Forrest M. Mims III, Kent Mok, Marshall Piros, Charles Platt, Marek Poliks, Alberto Sanchez, Becky Stern, Doug Stowe, Cy Tymony, Lee Wilkins

CONTRIBUTING ARTISTS
Mark Madeo, Rob Nance

MAKE.CO

ENGINEERING MANAGER
Alicia Williams

WEB APPLICATION DEVELOPER
Rio Roth-Barreiro

DESIGN

CREATIVE DIRECTOR
Juliann Brown

BOOKS

BOOKS EDITOR
Kevin Toyama
books@make.co

GLOBAL MAKER FAIRE

MANAGING DIRECTOR, GLOBAL MAKER FAIRE
Katie D. Kunde

GLOBAL LICENSING
Jennifer Blakeslee

MARKETING

DIRECTOR OF MARKETING
Gillian Mutti

PROGRAM COORDINATOR
Jamie Agius

OPERATIONS

ADMINISTRATIVE MANAGER
Cathy Shanahan

ACCOUNTING MANAGER
Kelly Marshall

OPERATIONS MANAGER & MAKER SHED
Rob Bullington

LOGISTICS COORDINATOR
Phil Muelrath

PUBLISHED BY

MAKE COMMUNITY, LLC
Dale Dougherty

Comments may be sent to:
editor@makezine.com

Visit us online:
make.co

Follow us:
🐦 @make @makerfaire @makershed
f makemagazine
▣ makemagazine
▶ makemagazine
ⓟ makemagazine

Manage your account online, including change of address: makezine.com/account
For telephone service call 847-559-7395 between the hours of 8am and 4:30pm CST.
Fax: 847-564-9453.
Email: make@omeda.com

Make: Community

Support for the publication of *Make:* magazine is made possible in part by the members of Make: Community. Join us at make.co.

CONTRIBUTORS

Who's your favorite fictional spy?

Gareth Halfacree
Bradford, United Kingdom (DIY Silicon!)
James Bolivar diGriz, the Stainless Steel Rat. A pill that sobers you up instantly sounds both useful and painful.

Laurel Cummings
Miami, Florida (Disarm a Drone)
Chuck Bartowski, because who doesn't love a tech nerd with an entire supercomputer in their brain?

Cy Tymony
Los Angeles, California (Hack Your Toothbrush)
TV's first Secret Service agent, James West. He battled colorful villains with 1870s gadgets, style, and unmatched physicality.

Issue No. 87, Winter 2024. *Make:* (ISSN 1556-2336) is published quarterly by Make Community, LLC, in the months of February, May, Aug, and Nov. Make: Community is located at 150 Todd Road, Suite 100, Santa Rosa, CA 95407. SUBSCRIPTIONS: Send all subscription requests to *Make:*, P.O. Box 566, Lincolnshire, IL 60069 or subscribe online at makezine.com/subscribe or via phone at (866) 289-8847 (U.S. and Canada); all other countries call (818) 487-2037. Subscriptions are available for $34.99 for 1 year (4 issues) in the United States; in Canada: $43.99 USD; all other countries: $49.99 USD. Periodicals Postage Paid at San Francisco, CA, and at additional mailing offices. POSTMASTER: Send address changes to *Make:*, P.O. Box 566, Lincolnshire, IL 60069. Canada Post Publications Mail Agreement Number 41129568.

PRINTED WITH SOY INK

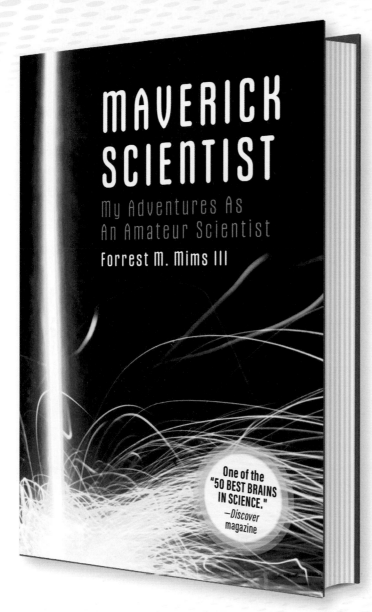

FROM THE EDITOR'S DESK

SKYNET FOR YOUNG PEOPLE

I received a message about your new book *Make: AI Robots*. I love the enthusiasm, but I'm concerned that discussion of ethics and AI's impact on society is missing from your editorial outlook. This is especially relevant for young people such as STEM students. What about the Three Laws of Robotics, Skynet, the displacement of knowledge worker jobs by Chat-GPT? It sounds like an old fart concern, but it's still relevant to the technical side of AI. —*Doug Crook, Salem, Oregon*

Editor Keith Hammond replies:
Thanks for the thoughtful note, Doug. We strive to explain technology's up- and downsides — see our discussion of Artists Against AI in "Generative AI for Makers" (Volume 84), "Face Jam: Evade Facial Recognition" (Volume 72), and this issue's Q&A with artist Dries Depoorter, who probes the darker aspects of AI and surveillance (page 42).

A NOVEL ECLIPSE VIEWER

I used this during the last eclipse near Kentucky. Google the pinhole camera equation for 30 feet; the aperture size is about ½" or 12mm. Make the aperture and place it over a mirror, reflecting the image back out to a screen; at 30 feet it is focused for safe viewing. You can cut a paper mask, or use a nickel as a guide to black out the rest of the mirror with a Sharpie marker. You now have a pinhole Newtonian reflector solar telescope.
—*David Fannin, Lexington, Kentucky*

PULL ME UP, PULL ME DOWN

Page 97 of **"Squishy Tech"** (Volume 86) says to use `pinMode INPUT_PULLDOWN` if you're using a modern Arduino-based microcontroller. There is no such mode of the `pinMode` statement for an Arduino. —*Don Lutz, Weymouth, Massachusetts*

Author Lee Wilkins replies: Right, the Arduino Uno doesn't have pulldown, but the microcontroller I specified in the repo, Adafruit Gemma M0, does. Many Arduino compatibles have pulldown; if yours doesn't you can use a pulldown resistor.

MAKER FAIRES MAKE ENGINEERS

Never underestimate how much your support did for the cohort of kids I worked with. Thanks to their involvement with Maker Faire, at least five of them went on to engineering and advanced degrees. You showed them what was possible.
—*Dave Lewis, Santa Clara, California*

ON THE BLOG: RIGHT TO REPAIR

Following our cover story **"War on Repair"** (Volume 80), Apple finally endorsed California's Right to Repair law, the nation's strongest, which passed the legislature unanimously. Kyle Wiens reports at makezine.com/go/apple-endorses-r2r.

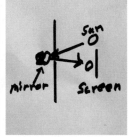

David Fannin, iFixit

GOT A PROJECT THAT BELONGS IN *MAKE:*?
Submit an article idea at make.co/submit-an-article-or-book-idea. See a succes on page 58!

Make:

THE BEST MAGAZINE FOR **MAKERS**, **DIY ENTHUSIASTS**, **HACKERS**, AND **TINKERERS**. SUBSCRIBE NOW!

makezine.com/subscribe

IN PRINT

ON THE GO

ONLINE

Reaping the Fields of Ingenuity

by *David J. Groom*, Community Editor at Make:

Last year's Boards Guide opened with an anxious yet hopeful description of the global chip shortage's impact on the automotive industry here in Detroit and on makers everywhere — but ended on an optimistic note about how makers were adapting and evolving in that challenging environment.

What a difference a year has made. Almost 200 boards were vying for a spot in our new 2024 guide, proving a tremendous challenge to fit just 81 into our special insert, and an impossible feat to choose just a dozen "new and notables" — observant readers will notice I cheated a bit and there's actually 14!

The first two on that list are highly anticipated new flagships from Arduino and Raspberry Pi — exciting progressions from previous generations, with magnificent new features to make our making merrier. Other featured boards range from tiny to titanic, highly specialized to hyper-performance, with all manner of adaptations like built-in cameras and displays, new silicon, and new ways to connect and expand our projects. Makers have patiently waited as worldwide supplies caught back up with demand, and their patience has been rewarded with a bevy of boards bountifully buyable from our favorite maker marts.

As we put finishing touches on this issue of the magazine, we are hurriedly preparing for the return of Maker Faire Bay Area where we'll share our excitement about these new boards at talks with Eben Upton and others, as well as at the *Make:* booth where we editors will have hardware on hand to nerd out over. Then it's on to Maker Faire Rome, and a visit to Turin with Massimo Banzi to find out what's cooking at Arduino. And then Maker Faire Shenzhen, hosted by Chinese innovators Seeed.

We're so excited for this new era of maker merriment. Boards are back, Maker Faires are back — and we can't wait to share it all with you in person and in these pages, and see how these seeds of ingenuity flourish and grow into next year's featured projects and exhibits! ⊘

STATEMENT OF OWNERSHIP, MANAGEMENT AND CIRCULATION (required by Act of August 12, 1970: Section 3685, Title 39, United States Code). 1. MAKE Magazine 2. (ISSN: 1556-2336) 3. Filing date: 10/1/2023. 4. Issue frequency: Quarterly. 5. Number of issues published annually:4. 6. The annual subscription price is 34.99. 7. Complete mailing address of known office of publication: Make Community, LLC 150 Todd Road Ste. 100, Santa Rosa, CA 95407. 8. Complete mailing address of headquarters or general business office of publisher: Make Community, LLC 150 Todd Road Ste. 100, Santa Rosa, CA 95407. 9. Full names and complete mailing addresses of publisher, editor, and managing editor. Publisher, Dale Dougherty , Make Community, LLC, 150 Todd Road Ste. 100, Editor, Keith Hammond, Make Community, LLC, 150 Todd Road Ste. 100, Santa Rosa, CA 95407, Managing Editor, N/A, Make Community, LLC, 150 Todd Road Ste. 100, Santa Rosa, CA 95407. 10. Owner: Make Community, LLC; 150 Todd Road Ste. 100, Santa Rosa, CA 95407. 11. Known bondholders, mortgages, and other security holders owning or holding 1 percent of more of total amount of bonds, mortgages or other securities: None. 12. Tax status: Has Not Changed During Preceding 12 Months. 13. Publisher title: MAKE Magazine. 14. Issue date for circulation data below: Fall 2023. 15. The extent and nature of circulation: A. Total number of copies printed (Net press run). Average number of copies each issue during preceding 12 months:48,033. Actual number of copies of single issue published nearest to filing date: 45,767. B. Paid circulation. 1. Mailed outside-county paid subscriptions. Average number of copies each issue during the preceding 12 months: 32,075. Actual number of copies of single issue published nearest to filing date: 30,350. 2. Mailed in-county paid subscriptions. Average number of copies each issue during the preceding 12 months: 0. Actual number of copies of single issue published nearest to filing date:0. 3. Sales through dealers and carriers, street vendors and counter sales. Average number of copies each issue during the preceding 12 months: 4,175. Actual number of copies of single issue published nearest to filing date: 3,366. 4. Paid distribution through other classes mailed through the USPS. Average number of copies each issue during the preceding 12 months: 0 . Actual number of copies of single issue published nearest to filing date: 0. C. Total paid distribution. Average number of copies each issue during preceding 12 months: 36,250. Actual number of copies of single issue published nearest to filing date;33,716. D. Free or nominal rate distribution (by mail and outside mail). 1. Free or nominal Outside-County. Average number of copies each issue during the preceding 12 months:505. Number of copies of single issue published nearest to filing date: 640 . 2. Free or nominal rate in-county copies. Average number of copies each issue during the preceding 12 months: 0 . Number of copies of single issue published nearest to filing date: 0. 3. Free or nominal rate copies mailed at other Classes through the USPS. Average number of copies each issue during preceding 12 months: 0. Number of copies of single issue published nearest to filing date: 0 . 4. Free or nominal rate distribution outside the mail. Average number of copies each issue during preceding 12 months: 788 . Number of copies of single issue published nearest to filing date: 905 . E. Total free or nominal rate distribution. Average number of copies each issue during preceding 12 months: 1,293 . Actual number of copies of single issue published nearest to filing date: 1,545 . F. Total free distribution (sum of 15c and 15e). Average number of copies each issue during preceding 12 months: 37,543 . Actual number of copies of single issue published nearest to filing date: 35,261 . G. Copies not Distributed. Average number of copies each issue during preceding 12 months: 10,490 . Actual number of copies of single issue published nearest to filing date: 10,506 . H. Total (sum of 15f and 15g). Average number of copies each issue during preceding 12 months: 48,033 . Actual number of copies of single issue published nearest to filing: 45,767. I. Percent paid. Average percent of copies paid for the preceding 12 months: 96.56% Actual percent of copies paid for the preceding 12 months: 95.62% 16. Electronic Copy Circulation: A. Paid Electronic Copies. Average number of copies each issue during preceding 12 months: 3,602. Actual number of copies of single issue published nearest to filing date: 3,570. B. Total Paid Print Copies (Line 15c) + Paid Electronic Copies (Line 16a). Average number of copies each issue during preceding 12 months: 39,852. Actual number of copies of single issue published nearest to filing date: 37,286. C. Total Print Distribution (Line 15f) + Paid Electronic Copies (Line 16a). Average number of copies each issue during preceding 12 months: 41,145. Actual number of copies of single issue published nearest to filing date: 38,831. D. Percent Paid (Both Print & Electronic Copies) (16b divided by 16c x 100). Average percent of copies paid during preceding 12 months: 96.86%. Actual percentage of copies paid for single issue published nearest to filing date: 96.02%. I certify that 50% of all distributed copies (electronic and print) are paid above nominal price: Yes. Report circulation on PS Form 3526-X worksheet 17. Publication of statement of ownership will be printed in the Winter 2023 issue of the publication. 18. Signature and title of editor, publisher, business manager, or owner: Todd Sotkiewicz - Business Manager. I certify that all information furnished on this form is true and complete. I understand that anyone who furnishes false or misleading information on this form or who omits material or information requested on the form may be subject to criminal sanction and civil actions.

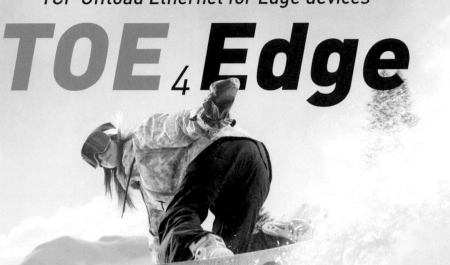

TCP Offload Ethernet for Edge devices

TOE₄Edge

Fast **Stable** $1.89/unit, 10K pcs

CoreMark Score

TOE Chip ████████████████ 173.34
SW TCP/IP ███████ 84.9

Tested on a Cortex M0+ by continuously sending and receiving
TCP data from an echo server

Iperf performance (Mbps)

W6100 ██████████ 50
SW TCP/IP █████ 25

Tested by Iperf at SPI 60Mhz

TOE Chip Applications

Energy Storage	28%	
Others	11%	
Smart Lighting	1%	
Laser Radar	2%	
Smart Traffic	2%	
Gateway	2%	
Audio & Video	3%	
Mini Printer	3%	
Bitcoin Mining	4%	
EV Charger	4%	
Industrial Control	6%	
Security Monitoring	8%	
Smart Grid	26%	

As of 2022

W6100-EVB-Pico

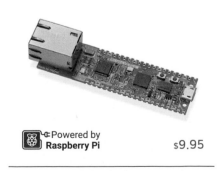

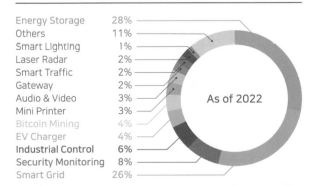

Powered by
Raspberry Pi $9.95

MADE ON EARTH

Amazing builds from around the globe

Know a project that would be perfect for Made on Earth?
Let us know: *editor@makezine.com*

BULLSEYE MACHINE
QUINTBUILDS.COM

Knife throwing is both an art and a skill that's difficult to master. It takes hand-eye coordination and just the right spin to embed the knife into a target, and the longer the distance, the more power and spin are needed. While action heroes make it look easy, for Oregon-based manufacturing engineer **Quint Crispin**, it was a challenge. Taking inspiration from the movies, Crispin and his son spent eight months building a machine that could throw knives and stick them accurately in a target at any distance within the machine's range.

Crispin designed his machine using a series of calculations concerning the force required to launch the knife, its mass, and of course its spin. Following this math, a pair of high-performance servomotors drive two parallel timing belts mounted to an aluminum frame. The belts then propel a carriage, with repurposed carbon-motor brushes holding the knife, along powered copper rails. One belt runs slightly faster than the other, which transfers to the carriage, creating spin as the knife is launched.

A solenoid-based mechanism secures the blades, which were modified with machined slots used for anchor points. The knives are fed automatically using a spring-loaded magazine housed in the back of the machine, similar to a bullpup rifle, and continuously when the trigger is held down.

The spin rate and distance needed to match that rotation are derived with the help of an onboard lidar, which also adjusts an aiming laser that paints the knives' intended location on a target. Crispin's son **Shane** handled the programming and coded the lidar sensor via I²C using the C programming language.

As with any project, there were revisions. Unstable knife retention systems, explosive carriages, and faulty electronics plagued the build; however the end result speaks for itself.

While the Knife Throwing Machine has no practical applications, it's a great example of what can be created in the mind of an engineer.
—*Cabe Atwell*

Orissa Crispin

ALL-ELECTRIC 18WD OFF-ROADING

18WHEELS.FI

Most land-based vehicles are not particularly unique in their designs, specifically regarding their number of wheels. Bicycles have two, tricycles have three, cars have four, and big rigs have even more. However, an innovative Finnish company aptly named 18 Wheels reimagined this sacred vehicle-to-wheel ratio with an electric ATV design featuring — you guessed it — a fully functioning 18-wheel drive.

The company's founder, **Eldar Aliev**, has been closely following the evolution of electric vehicles for many years and is well-versed in the untapped potential of the electric hub motor or wheel motor — and its undeniable problems. Aliev explains, "Although it's evident that the wheel motor is the optimal solution for electric transportation, their use significantly increases the unsprung mass of the vehicle ... especially in high-speed driving." While working on alleviating these weight distribution difficulties and motor efficiency issues, Aliev and his team hit upon a creative solution that involved decreasing the wheel's diameter. But because a small wheel can't handle diverse terrain alone, 18 Wheels invented a suspension system capable of supporting nine axles within the dimensions of a standard ATV or snowmobile — at less than half its original weight. The resulting creation looks like a gigantic centipede crawling across the ground, albeit one that can smoothly traverse virtually any surface at almost 40 miles per hour.

So many wheels may seem like a strange choice, but it's all calculated according to Aliev's design. "Because our ATV has 4.5 times as many wheels as a typical car, we were able to reduce the power of each wheel's motor by 4.5 times without sacrificing the rest of the vehicle's power, which ultimately results in a 22.5-fold increase in the electric motor's efficiency." The company is currently building its second prototype within a snowmobile-like frame, and they plan to apply its creative solution to other vehicle classes in the future. What is Aliev's ultimate goal with this idea? "The patent for the electric wheel motor was filed nearly 130 years ago, yet its widespread application has been lacking; we believe a renaissance of wheel motors is possible with our technology." —*Marshall Piros*

Ilia Timoshkov

A PERFECT FIT

YOUTUBE.COM/@HANDTOOLRESCUE

Imagine: a wooden chair that comfortably conforms to every contour of your unique buttocks, *without pinching them*.

We introduce to you the world's first "fractal" chair — based on a potential application of an expired 1913 patent, invented by Paulin Karl Kunze, for a "fractal vise," designed for clamping bodies of any shape. Canadian maker **Eric Tozzi** went viral two years ago for restoring the rare antique device on his popular *Hand Tool Rescue* YouTube channel, attracting over 21.3 million views. It laid the foundation for this 150-pound chair that, to Tozzi's knowledge, has never previously been made.

"I thought that if you based it somewhat off the drawing, it could be done now using modern techniques," he tells *Make:*, explaining he partnered with Fick Tool & Design to create a 3D design mockup, and then outsourced to a fiber laser company to cut the metal parts.

The chair has been over two years in the making, with most of the time spent on conceptual planning, and then about one month of "fiddling" with the parts, as Tozzi says, in the evenings after working as a plant scientist at the University of Saskatchewan.

Despite the impressive creations he's been showcasing on YouTube since 2017, Tozzi clarifies, "I am in no way an expert in chair design, nor an expert machinist or woodworker on any professional level — just incredibly enthusiastic about these types of projects bringing patents back to life."

That enthusiasm carried him through the biggest challenge: very careful machining of 30 bars supporting the apparatus to ensure functionality. "If they're off by even a few thousandths of an inch, the actual mechanisms of each fraction will jam and not move at all," he says. "So, it needs to be incredibly high-precision machining, which takes a really long time." —*Greg Gilman*

Hand Tool Rescue

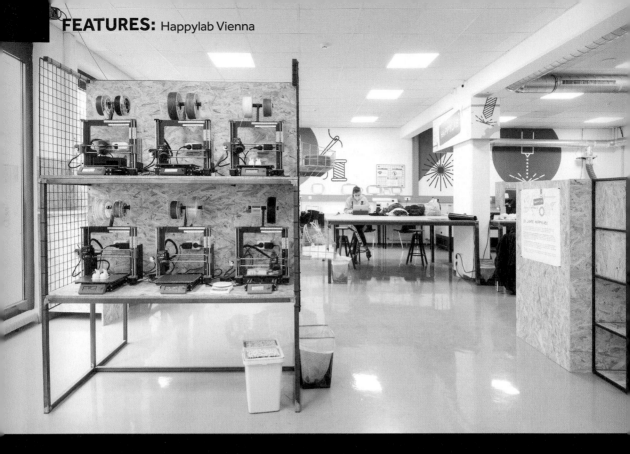

A MODEL MODERN MAKER SPACE

Vienna's thriving **Happylab** is helping other makerspaces succeed

Written by Dale Dougherty

DALE DOUGHERTY is the founder of *Make:* and the president of Make: Community.

The founders of Happylab met while working on the autonomous sailboat *ASV Roboat*, shown here off the coast of Portugal en route to victory in the World Robotic Sailing Championship in 2009.

Happylab founders and co-CEOs (left to right) Roland Stelzer and Karim Jafarmadar at Happylab Vienna in 2023.

© Lukas Bast, Dale Dougherty

Happylab Vienna, the first makerspace in Austria, was inspired by Professor Herbert "Happy" Hörtlehner, who taught electronics and robotics at local schools and at the University of Vienna. Both of Happylab's founders, Roland Stelzer and Karim Jafarmadar, were students of the professor in computer science, although at different times.

"Happy had a very special way of teaching," Karim told me during an interview at Happylab. "He didn't come into class with any plan, but he asked 'What do you want to do?' And then we learned kind of by just doing and trying things out. It was kind of the maker way to learn things."

The professor's apartment was a place where his students past and present gathered for tinkering and sharing projects. While studying at university, Karim was part of a project to build a robotic sailboat and the team met at Happy's apartment. Roland met Karim there and they both worked on *ASV Roboat*, which eventually won the World Robotic Sailing Championships four times in a row. At the start, *Roboat* was a small model boat, 1.5 meters long. It could sail autonomously and then find its way back.

In 2005, Happy Hörtlehner passed away suddenly from a heart attack while on vacation. His students no longer had that place to meet outside of school and learn about new projects that others were doing.

DER MAKERS-KELLER

Karim and Roland continued to work on the *Roboat* project and even bought a bigger sailboat, 4 meters long. On the way home with the new boat, Karim started thinking about where to put it.

They found some space to rent cheaply in a cellar in the neighborhood. "We couldn't get the boat in actually, because it was too big," he said. So they ended up having to rent a garage as well.

They hadn't intended to open the cellar for others. But as soon as they had the space, other people wanted to use it. In 2006 the cellar became the first Happylab, an open workshop for tech enthusiasts, filling the need for a drop-in space that their professor's apartment had once filled.

"We thought that it was cool that people came by," said Karim. "We're nice people and we didn't

Workbench in the electronics area.

Welding station in the metal workshop.

Wood workshop with large-format CNC milling machine for professional woodwork.

The innovation workshop for startups, founders, and makers.

Workshop with hand tools.

3D printer (fused layer process).

Laser cutter for metals.

Working on the computer-controlled CNC milling machine.

need all the things we had, all the time. So we opened it up. People just came and continued coming."

SEEDS OF A MAKER COMMUNITY

A small community started growing. "We thought that the people who would come would be technology nerds," Roland recalled. "But what we saw very soon is that it was much wider —artists, people who were just curious about 3D printers, craftsmen, and more. So it was almost anybody."

Schools and universities might have the same equipment, even better equipment than Happylab, said Roland. "The special thing about our space is the people who come together and meet each other, who help support each other and work on projects together."

Happylab reflected a mindset that anyone could learn to do things with some basic help. "This is probably something we learned from the Happy professor," said Roland. When they added a CNC machine, they created a one-hour workshop so that anyone can learn to use it. "You don't need to have any background knowledge. After one hour, you can learn how to do simple things with the machine," said Roland. "You should learn enough to not hurt yourself and not damage the machine. The rest is learn-by-doing and watching others."

Happylab outgrew the original cellar and moved to a larger 2,500-square-foot space in 2010. "Every time we moved was because we could not fit any more people or machines into the space. It was the next logical step," said Karim. Each move came with risks, which they took on, even though it was basically the two of them responsible for paying a higher rent. "So far, each time it has worked out," said Karim, as more people came to use the larger space.

In 2021 Happylab moved to their current 10,000-square-foot space in the Stuwerviertel neighborhood near Vienna's famous Prater park. By that time, they had opened new locations in Salzburg (2014-2020) and Berlin (2016–present).

AUTOMATING A MAKERSPACE

As both of the founders were programmers, they wanted to automate as much as possible the management of a makerspace. "We don't want to

> **"WE THOUGHT THE PEOPLE WHO WOULD COME WOULD BE TECHNOLOGY NERDS, BUT IT WAS MUCH WIDER — ARTISTS, PEOPLE WHO WERE JUST CURIOUS ABOUT 3D PRINTERS, CRAFTSMEN, AND MORE. SO IT WAS ALMOST ANYBODY."**

Happylab designed the Fabman automation system for makerspace management.

have people having to do all the boring things that have to be done," said Karim. They automated access control to the building for members, who can come at any time of day or night. Automation freed them up to focus on workshops and community building. It also helped them grow membership to 1,600 members.

But it was a 2010 incident shortly after opening their second space that made Happylab really focus on safety and unsupervised users. "We had just moved in and everything was new and we were proud of it," said Roland. "One of our power users did some laser cutting and the job was almost finished, but he was not completely done when he decided to go to lunch. When he came back, six fire engines were outside the building and the new lab was on fire."

Roland wrote about what happened on their website: "The laser cutter's axis had gotten stuck, causing the laser to work on the same spot for a prolonged period. The plexiglass [workpiece] caught fire, followed by the laser cutter's lid, followed by the whole ceiling. My co-founder Karim was in the lab's office when it happened. He only noticed the fire when the power went out because the electrical wiring in the wall was starting to melt."

"That's not the very best start," said Roland. They closed for two months and had to renovate everything. Luckily, insurance covered most of the costs. Roland and Karim realized they needed a solution to prevent this kind of problem happening again. They laser-cut a wooden case for an Arduino and a chip card reader and connected it to one of their fabrication machines. They called the device a bridge. "Now, you have to authenticate yourself before you can switch on the machine and you have to stay at the machine for five minutes, or you have to show your card again," Roland explained. "Otherwise it goes into an emergency shutdown and stops working."

Once they'd added card readers for each machine, they thought the system could be used for booking time on machines as well as charging for usage. "When makers from other labs visited us, they said that 'We are facing exactly the same problems and that's what we need,'" said Roland. Eventually, Happylab's solution for managing a makerspace became a product called Fabman. However, turning their one-site solution into one that worked for many different kinds of makerspaces required them to start from scratch and build a new, well-engineered system. It's a software-as-service that operates from a web browser. Over 120 makerspaces around the world now are using Fabman, from small private spaces to university labs.

"We always say it's the operating system for a lab," said Karim. "So this is something you get; it should always work because this is the most important part. And then you can build your own applications on top of it."

Each makerspace that uses Fabman generates usage statistics, which could be a valuable way to understand many aspects of a makerspace.

A stuck gantry and absent operator caused this laser cutter to ignite the workpiece, itself, and Happylab's brand new space in 2010.

Roland said that "so far, 443,075 machining jobs from a total of 4,927 different users have been logged in Happylab's Fabman system. In total, that's 270,934 hours of making."

COINS AND MINUTES

Fabman also helps address a problem for membership-based spaces that charge casual users and heavy users the same monthly fee. That fee might cover some basic machine usage and heavy users could be asked to pay more based on their usage. Fees could also be higher during times of peak demand.

"There are those who are hobbyists and those who are professionals," explained Roland. "When we charged a higher flat fee for professionals, many of them said they were hobbyists and tried to hide the work from us, just to get the lower monthly fee." Eventually, Roland and Karim decided not to distinguish between the two groups anymore. They introduced a credit system with "coins." With the monthly fee, each user gets a certain amount of coins. When the coins are used up, then users pay for the additional minutes they use. "I think it's a fair way," said Roland. "It allows us to offer monthly usage at a reasonable price."

Happylab moved into their current space in Vienna during Covid-19. It's about twice as large as their previous space, with a large main room with work tables as well as 3D printers and laser cutters. There's a classroom that's used for workshops, and dedicated rooms for metalworking, woodworking, ceramics, and a materials shop.

Sometimes Roland wishes their makerspace was one big room with all their members and machines visible in one space, but he knows that's not practical, as some machines are loud and some are more dangerous than others. "What works," he says, "is having a large common area with a big table in the middle, so people can see what the others are working on."

FIFTH MAKER FAIRE VIENNA

When I met with them in June, Roland and Karim had just finished organizing their fifth Maker Faire Vienna after a three-year hiatus due to Covid. The event was held in a two-story industrial building that once was used for building steam engines. Maker Faire Vienna had about 300 makers exhibiting for a crowd of about 10,000 people. In Happylab's booth, they displayed the work of some of their makers. MiniMotoz (minimotoz.at) are CNC-cut and laser-engraved kids' balance bikes built by Robert Poeckh. Epiphany is a nightstand for capturing ideas when you wake up, by Benedict Heinzl and Dominik Glatzl of GoldenRatio.

I asked Roland how the public perception of Happylab has evolved over time. "It definitely changed in a positive way," he said. "In the beginning, some saw us as nerds playing around with crappy tools but this has definitely changed." They work closely with the Vienna Business Agency, which also provided financial support for the setup of the new space. "There are more and more success stories," said Roland. "I always see it like a football team in a small village, where every young boy or girl can enter and can start playing football there. You don't need any prior knowledge, but when you see some talent, you have to guide them to the right people, to the right tools, to the right knowledge; they need skills and help to go on to the next level."

Having managed a makerspace for nearly 20 years, how have they avoided getting burned out? "At least for me personally, it's important that there are two of us," said Roland. "You can't always be 100 percent motivated so you need to have a good team around you."

Karim said, honestly, that he doesn't always like having two people in charge. "We discuss everything like thousands of times until we come to a decision," he said. "So even though we're just a small company, it takes us ages to decide on something." Nonetheless, he believes that together they make better decisions because they

Happylab booth at the fifth Maker Faire Vienna.

MiniMotoz wooden balance bikes, prototyped at Happylab Vienna.

Dale Dougherty

(Left to right) Dale Dougherty with Happylab founders Roland Stelzer and Karim Jafarmadar, community manager Vjara Jovkova, and members Lukas Winter and Martin Unterberger.

have thought about it so much.

The duo and their team have managed to build and sustain a successful community-based makerspace with more than 2,000 members from Vienna to Berlin, which they believe makes Happylab "the largest maker community in Europe." Others are trying to replicate their model, and that pleases Karim. "It is a nice recognition from others that they tried the same way because they think this is a good way to do it," he said.

"Happylab was an experiment from the beginning," Roland added, "and still is." ⊘

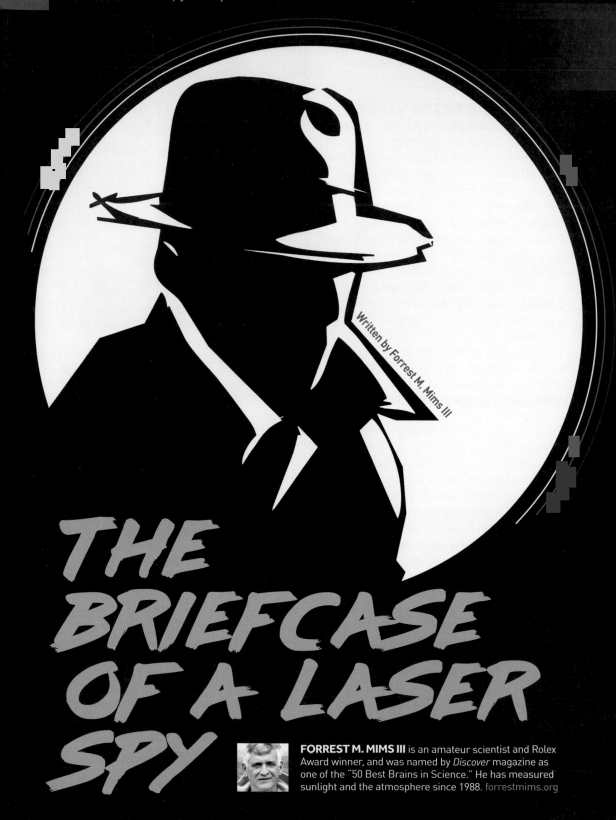

Written by Forrest M. Mims III

THE BRIEFCASE OF A LASER SPY

FORREST M. MIMS III is an amateur scientist and Rolex Award winner, and was named by *Discover* magazine as one of the "50 Best Brains in Science." He has measured sunlight and the atmosphere since 1988. forrestmims.org

MY LASER EAVESDROPPING ADVENTURES WITH THE NATIONAL ENQUIRER

There was a time when I was intrigued by the technology of spying, especially the incredible devices developed for Agent 007 by Q in the James Bond movies. While Q was explaining to Bond how to use his latest fictional gadget, I was trying to figure out how to make a working version. I did not want to be a traditional spy; I just wanted to design their instruments.

When I was growing up, I spent hours experimenting with invisible ink and devising secret codes. When I was a senior at Texas A&M, a CIA recruiter learned about my miniature radios and light-wave communicators and encouraged me to apply for an intelligence position.

I took a different path in the Air Force, but an interesting spying opportunity arose in 1975, when a letter arrived from the *National Enquirer* to my book publisher Howard W. Sams and Co. seeking my advice for an article "outlining the growing role of lasers in the everyday life of the people of America."

The *National Enquirer* had a terrible reputation, but I was curious about what seemed to be a serious journalism project, so I called Enquirer editor Bernard Scott and mentioned twelve topics related to lasers. Scott asked whether I was aware of reports that a laser could intercept conversations in a closed room by detecting the reflection of the laser beam from a window. Voices in the room would cause the window to vibrate in step with the sound, thereby imposing the voices as amplitude variations in the reflected laser beam.

Scott was surprised when I told him I could build a laser apparatus for this purpose. He then shifted the conversation to billionaire Howard Hughes, whose private life the *National Enquirer* was investigating to uncover what Scott described as suspicious business practices. Would I be willing to assist by building a laser bugging device? When I said maybe, he asked me to send him a proposal. He also strongly emphasized that our discussions should be kept secret.

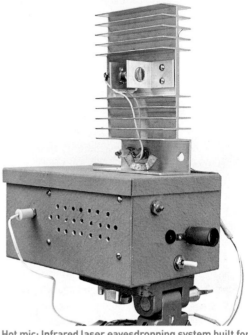

Hot mic: Infrared laser eavesdropping system built for the *National Enquirer* in 1976.

THE LASER EAVESDROPPING PROPOSAL

I quickly sent an eight-page proposal that described twenty laser applications that could be covered in a major article to be titled "The Laser Comes of Age" or "There's a Laser in Your Life." The laser-bugging proposal Scott requested was also included. I titled it "Listening in on a Personality."

On September 5, Scott called. He ignored the article proposal but was highly interested in the laser eavesdropping proposal. He said the plan was for me to build the apparatus, fly it and me to Florida, and impress the publisher by using it to intercept conversations in the publisher's office. He provided a detailed description of the office, including its dimensions, the windows, and the location of the desk. Scott said that "this capability" would provide an important method for the *National Enquirer* to verify some of its major stories. He emphasized that the project must be considered "top secret" — a phrase I had not heard since working in the Air Force Weapons Lab.

Days later, I called Scott to report my tests, in which the sound of a radio was detected by pointing the very narrow beam of a helium-neon laser at the window of the room in which the radio was playing. Scott asked me to make a tape recording of these experiments. He then said that "the device" should be reliable and easy to use "when it's needed." That's when it became clear they wanted to keep whatever I built — and that they were seriously planning to bug Howard Hughes.

I sent Scott the tape recording with a letter explaining my concern: "I do not want any involvement with illegal operations but see an exceptionally unusual article in exposing the security weaknesses in various government offices ... and describing our little technique for the public."

Scott responded positively and said the *National Enquirer* had contacts who were close to Sen. Barry Goldwater. He suggested we might get permission to spy on Goldwater's office in the Senate Office Building to demonstrate the security problem posed by Soviet laser eavesdropping.

I continued to work on the laser apparatus and prepared a detailed progress report for Scott. On November 12, he agreed in writing that the *National Enquirer* would pay for the purchase of a telescope and precision micrometer for the project for up to $350 ($1,890 in 2023 dollars). He also agreed to pay a weekly rate for my time. I explained that while the red helium-neon laser worked well, it was so bright, it would be easily seen at night. Therefore, I needed to assemble a near-infrared laser illuminator that emitted an invisible beam.

The optical supplies arrived in late December, and on January 8, 1976, I wrote Scott that a battery-powered laser system with an invisible beam was completed. I then began work on a receiver system having much more sensitivity than the one used previously. On January 22, I called Scott to inform him that all the equipment was finally assembled.

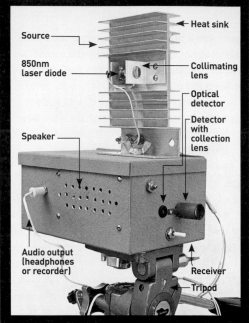

LASER EAVESDROPPING SYSTEM FOR *NATIONAL ENQUIRER*, 1976

Labels: Source · Heat sink · 850nm laser diode · Collimating lens · Optical detector · Detector with collection lens · Speaker · Audio output (headphones or recorder) · Receiver · Tripod

The **laser source** and **receiver** were mounted together to simplify operation in the field. What was then a state-of-the-art, continuously operating **near-infrared diode laser** (Laser Diode Labs LCW-10) was powered by a **6V battery** through a **current-limiting resistor** that protected the laser by limiting its emission to 6mW. The invisible (850nm) emission from the laser was focused into a tightly collimated beam by a small **f1 lens** mounted on the large **heat sink** that dissipated heat from the laser.

The receiver's detector employed a **phototransistor** with or without a **lens**, for the reflected laser beam was very small. The phototransistor was connected to a **FET op amp** having a peak amplification of 100,000, followed by a **low-pass filter** that removed noise from the laser and stressed the frequencies of human voice, and finally an **audio amplifier** connected to either a small **speaker** or an external **earphone**.

It was essential to mount the receiver on a sturdy **tripod**, wear dark glasses when looking at the target, and avoid pointing the system at people.

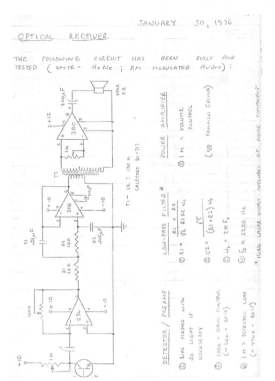

Tycoon target: Howard Hughes's penthouse suite at Xanadu Princess Resort in the Bahamas.

Q Branch engineer's notebook: Forrest Mims's original schematic diagram of the laser eavesdropping rig.

LASER BUGGING THE *NATIONAL ENQUIRER* HEADQUARTERS

On February 9, I arrived in West Palm Beach and assembled the laser apparatus in my briefcase at a Howard Johnson hotel. The following morning, Scott and reporter Tony Brenna drove me to the *National Enquirer*'s main building.

Because of the paper's sleazy reputation, I was surprised by the impeccable landscaping and the neatly organized desks and offices inside. Sensing my concern, Scott and Brenna claimed they were much more careful about fact-checking than other major newspapers. That was why they were so highly interested in laser eavesdropping, they explained, for they hoped it could be used to surreptitiously verify facts.

Before setting up the laser gear, I spent an hour at Scott's desk discussing the project. Had they contacted Senator Goldwater's staff about arranging a demonstration in Washington? But Scott had only one topic in mind: Could the laser equipment be used to spy on Howard Hughes? He

produced photos of the 13-story Xanadu Princess Resort & Marina in the Bahamas and pointed out Hughes's bedroom on the right side of the top floor, the parapet that shielded the top floor from the one below, and the pyramid-like structure atop the roof that blocked helicopters from landing there.

I explained how the laser system could be used only against a normal window and not one set at an angle. This meant that a laser beam pointed at Hughes's thirteenth-story bedroom window would be reflected toward the sky instead of the receiver on the ground. Scott then casually said that the *National Enquirer* would pay $100,000 (over $500,000 in 2023) for a full-face photo of Howard Hughes.

I suggested that they could fly a quiet, radio-controlled blimp equipped with a camera outside Hughes's window and take a photo when he looked outside. I was smiling, but Scott took me seriously and asked if I knew anyone with such a blimp. *American Modeler* magazine had published an article about such a craft several years before, I suggested. Scott immediately called an assistant to his desk, and within 15 minutes, they were speaking with the blimp's owner by telephone.

I was curious why they were so interested in Hughes and asked whether he had any connections with organized crime. Instead, they told me that the *National Enquirer*'s owner might have such connections, via his father. They said a quick call to the "right" person was occasionally needed to get the delivery trucks moving.

By midmorning, I had set up the laser

TIME back issues found on eBay

Mad money: Eccentric billionaire Howard Hughes became a national obsession in his reclusive later years.

apparatus on the carefully manicured yard outside the building, 100 or so feet (30m) from the glass window of a conference room in which we had placed a radio tuned to a talk program. As I began a series of tests, the equipment and I began to attract a growing crowd of *National Enquirer* staff who came outside to watch what they were told was a top-secret project. Occasionally I could see glances from a tall, distinguished-looking man when he stood up at his desk behind a window in the building. He was Generoso Pope Jr., publisher of the *National Enquirer*.

Back in Albuquerque, the laser system had worked reasonably well. But the *National Enquirer* demonstrations were temperamental. All the windows were made of extra-thick, hurricane-proof glass that refused to vibrate as well as the window glass I had tested back home. I successfully demonstrated the device by concealing a tiny laser reflector inside the conference room — but they wanted a system that did not require inserting anything inside a room being monitored.

That night, we tried again and were successful. When the window was 112 feet (34m) away from the laser and receiver, slight adjustments of the receiver lens within the reflected beam finally allowed conversations in the room to be intercepted. The next day, we had similar results using both the red helium-neon laser and the invisible-beam laser. All these tests required considerable time and careful adjustment of the position of the laser. This method might work if the laser system was installed in a van parked on a street opposite a window of interest. But that is not what the *National Enquirer* had in mind.

Scott gave me a check for $2,020.85 ($10,912 in 2023 dollars). This included $261.47 for the flight to West Palm Beach, electronics expenses, and $500 a week for 3½ weeks of work. The following morning, I took a walk along the beach to reflect on the events of the previous two days and washed it all away with a swim in the Atlantic.

Eight weeks later, Hughes would be dead.

LASER EAVESDROPPING REDUX

Later I wrote several articles about laser spying, including "Surreptitious Interception of Conversations with Lasers," the November 1985 cover story for *Optics & Photonics News* (doi.org/10.1364/ON.11.11.000006). The cover photo showed a miniature laser source and the receiver built for the *National Enquirer* project in a miniature briefcase over the headline "The Case of a Laser Spy."

Around the same time, John Horgan, a reporter for the Institute of Electrical and Electronics Engineers, published a full-page story about my *National Enquirer* adventures in the organization's tabloid, the *Institute*, after I gave a talk at an IEEE meeting.

These articles attracted attention in the media. HBO flew my son Eric and me to New York to stage a demonstration of laser eavesdropping between two apartment windows for one of their shows. I was also approached by an IRS agent to assist with an investigation of an organized-crime group in Chicago, and by a man who claimed to be with the FBI and asked for a confidential demonstration of the eavesdropping equipment, though neither followed through.

Finally, there was a computer convention in Los Angeles, where I received a mysterious request

Eyes on spies: (Above and top right) The cover story of *Optics News*, November 1985, showing a mini laser source with Mims's eavesdropping receiver from the *National Enquirer* job. Mims sought to warn U.S. officials that laser bugging was a serious security threat.

for a private meeting. After I was escorted to a curtained-off area, a remarkably beautiful woman dressed like a movie star arrived and earnestly pleaded for my help to prove that her ex-boyfriend was bugging her house with a laser. I explained the technical difficulties and how unlikely that was. We went back and forth for 30 minutes, but she was not convinced.

None of my projects have attracted such a diverse range of people as laser eavesdropping, and this especially includes John Horgan, who became an editor at *Scientific American* magazine. In 1989, I sent him my proposal to take over their column "The Amateur Scientist." Had I not known Horgan, my misadventure with that famous magazine might never have occurred, and I might never have begun doing serious science. ◗

Launch code: Mims with an original 1970 MITS rocket displayed at the New Mexico Museum of Natural History and Science exhibit "Startup: Albuquerque and the Personal Computer Revolution."

This article is excerpted from Forrest Mims' new memoir *Maverick Scientist: My Adventures as an Amateur Scientist*, now available at makershed.com/products/maverick-scientist-hardcover.

DISARM A DRONE

INVASIVE DRONES ARE SNOOPING EVERYWHERE — HERE'S HOW THEY CAN BE DEFEATED

Written by Laurel Cummings

LAUREL CUMMINGS is a freelance electrical engineer based in Miami, Florida, specializing in rapid prototyping and disaster response. She is also a flight coach for zero-gravity flights with Zero-G.

Drones have proliferated everywhere in recent years, from underneath Christmas trees to major sporting events and even warzones. They're cheap, easy to pilot, and have a heck of a lot more range than the Dollar Tree RC helicopters I grew up with.

But with that accessibility comes the reality that anyone can get their hands on a drone, and that drone can become a menace — spying with cameras or Wi-Fi sniffers, delivering hostile payloads, invading your space and your privacy.

In the U.S., there are laws meant to protect everyday folks from reckless drone operators, and by March 16, 2024, all FAA-registered drones will be required to broadcast their **Remote ID and location** — think of it as a license plate for drones, allowing the aircraft to be traced back to its pilot to ensure everyone is flying safely.

But there are still bad actors who don't adhere to the rules, and drones smaller than 250g can fly unregistered. So how can we disarm a troublesome drone? There are a number of ways a rogue drone can be neutralized or brought down, from RF exploits to good ol' fashioned "throw something at it" methods.

NETS

Sometimes a complex problem has a simple solution. One of the easiest and most consistent ways to bring down a drone is to disable the propellers, causing the drone to crash back down to Earth. The Tokyo Police Department's **interceptor drone**, for example, drags a net through the air to catch the rogue drone by forcing a mid-air collision with the net.

An extremely effective DIY method of catching low-flying drones is a simple **net gun**. Modify a T-shirt cannon or an old potato cannon to use a net instead of the usual projectile, and you've got yourself a DIY drone wrangler! The tricky part is designing a launcher that encourages the net to spread out when launched. Like in shrimping, the net must be cast in a spreading motion with weights at the ends, to catch a drone and stop those props. Try designing a 3D-printed capsule that helps disperse the net, or try Crispyjones'

AirWarden Remote ID receiver, from AeroDefense.

G-16 antenna for DJI's discontinued AeroScope drone detector.

net gun at instructables.com/Build-A-Net-Gun or William Osman's at youtu.be/Z5plajOgCzc.

LASERS

When we imagine how to protect ourselves from the threats of the future, the answer is usually an enthusiastic "lasers, of course!" But how effective are lasers at defending against drones?

A drone is susceptible to extreme heat, especially if its wiring or LiPo battery gets compromised. However, a laser powerful enough to physically damage a drone is at least a **Class 4 laser** and your aim would need to be very steady.

If you're determined to have a Death Star moment with the pesky drone buzzing your backyard, any laser you can get your hands on is more likely to mess with the drone's **cameras or IR sensors** rather than melt it. Remember that shining a laser beam at any aircraft, including drones, is very illegal in the United States and could land you with hefty fines or even jail time.

SENSORS

Most drones are loaded with sensors, from inertial measurement units (IMUs) to magnetometers to weather instruments like barometers for calculating altitude. Some of these sensors are considered the "payload" of the drone, recording data during flight for later analysis, but more often than not, these sensors feed into the flight controls of the drone, especially if it's programmed for an automated flight path, or flying in a stabilized mode.

For example, DJI's Mavic Mini drones have multi-direction **time-of-flight (TOF) distance sensors** that help prevent the drone from crashing into any solid objects. These sensors have an infrared (IR) emitter and receiver, sending out a pulse of infrared light, then measuring the

DIY pneumatic net gun by William Osman.

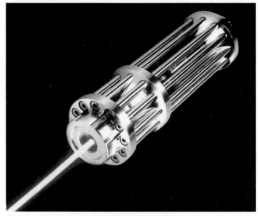

Class 4 blue laser from **c4lasers.com**.

Long-range IR illuminators, from Amazon.

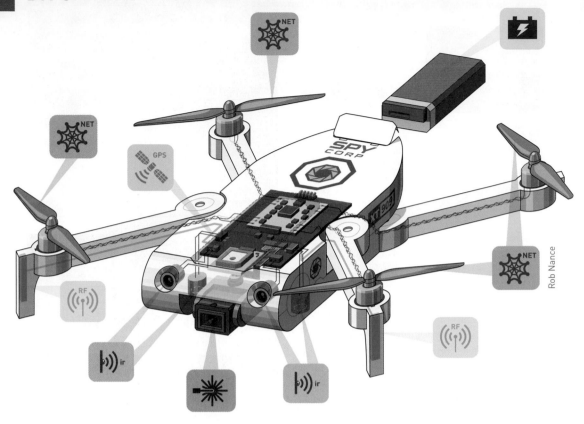

Rob Nance

time it takes for that light to be reflected back. Using that time, the sensor can calculate a precise distance from an object or surface, such as the ground or a wall. TOF sensors can also be used to 3D scan and map objects.

Of course, one drone's safety feature is another person's chance to exploit. With a bright enough **near-IR flashlight** to flood the sensor, or **refractive material** to scatter the light (don't fly a drone over a swimming pool!), it is possible to dupe the TOF sensors into believing the drone is about to collide with an object. Depending on how the flight controls are configured, you can bias the drone into overcorrecting to avoid a collision, like a technological matador!

LIPO BATTERY

Lithium polymer (LiPo) batteries are common to most drone platforms. They have all the right characteristics making them the perfect choice as the power supply for unmanned vehicles, with their high energy density, quick discharge rate, and light weight.

LiPo batteries, however, do have drawbacks —

like their propensity to **burst into flames when punctured**. If there's one surefire way to bring down a drone, it is causing said drone to light itself on fire! If you decide to take the medieval route to take down a drone using a **crossbow bolt or similar projectile**, be aware that if you puncture the LiPo battery, the drone may cause a fire where ever it crash-lands.

21st-century hunting compound crossbow.

GPS JAMMING

There are two paths when it comes to messing with a drone wirelessly — attack its GPS signal or its radio frequency (RF) control signals. To choose your avenue of attack, study the type of flight path the drone is taking. Drones can either be controlled via a live operator or be programmed to follow an automated flight path. If it's on a programmed path, it's relying on GPS.

GPS jamming is highly effective — but extremely illegal — for stopping rogue drones. When a drone loses its GPS fix, it will usually rely on other sensors to maintain altitude, but will drift in the air until the operator intervenes or the drone lands itself in place.

GPS satellite signals are notoriously weak when they reach the GPS receiver, making it surprisingly easy to jam the signal. There are three GPS frequency bands known as L1, L2, and L5, at 1575.42, 1227.60, and 1176.45MHz. GPS jammers can be bought online for very cheap, but the problem is broadcast strength. If the drone is too far away, you'll need a louder transmitter. Pick up an **RF amplifier** and a **directional antenna** to avoid jamming unintended targets.

GPS SPOOFING

GPS spoofing is a trickier art form, but produces much more exciting effects when used on a drone. A commonly available **software defined radio (SDR) transmitter** and some open source software is all you need to produce a false GPS signal. The drone will try to correct itself based on where it was and where it thinks it currently is. This can throw the drone into an over-correction loop, where the sensors and the GPS are at war with one another in the flight controller module. It produces an effect much like "death wobble" on a skateboard, where the oscillations build up until you're thrown off, or in the case of the drone, it's sent crashing into the ground!

RF JAMMING

GPS denial is a great trick in the anti-drone toolbag, but it won't lock a drone operator out of control of the drone entirely. Instead, jamming the radio control signals is a reliable way to stop delinquent drone operators from buzzing your yard and annoying your dog.

Most drones operate on the 2.4GHz and 5.8GHz radio frequency bands, utilizing Wi-Fi, Bluetooth, and other analog and digital broadcast methods to transceive control, instrumentation, and video data. So the "easy" solution is to drown everything out on those RF bands, right?

Well, unless you've gotten your hands on a very expensive and very loud radio transmitter,

HackRFOne software defined radio (SDR) transceiver, with PortaPack H2 add-on mobile interface.

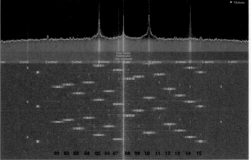

Frequency hopping sequence on 15 channels, captured in software defined radio (SDR) by Dr. Fraida Fund.

you'll need to go the clever route to jam a drone. Most drones utilize a technique called *frequency hopping,* which means they switch rapidly between several (usually eight) narrow slices of those RF bands, so they don't interfere with other devices and drones in the area, but they're still receiving the correct data.

Using an **SDR**, it's possible to listen and figure out which narrow bands the drone is hopping between. Then, you can set up multiple smaller (and cheaper!) **RF jammers** on those step frequencies, so you're jamming precisely, and thus need less power! It pays to work smarter not harder, especially in the realm of RF jamming.

RF HACKING

Really want to crack the code? An **SDR** can retransmit captured RF signals to cause confusion in a *replay attack*. Or, a **Wi-Fi** *deauth attack* can disconnect a drone from its controller. You could even try to decode the signals with **GNU Radio** and then send your own commands to **hack or spoof** the drone's onboard computer. Every drone's different, so you're on your own here. ●

Wikimedia/Public Domain, Amazon, Fraida Fund

HIDING IN PLAIN SIGHT

SEND SECRET MESSAGES INVISIBLY OVER LIGHT WAVES WITH YOUR OWN OPTICAL COVERT CHANNEL

Written and photographed by Joe Grand

Adobe Stock-Vector Tradition

The art of spycraft often requires using clever methods to send secret messages. This can be done with invisible ink, ciphers, dead drops, and even light!

This project is your entry into *covert channels* and the world of optical communications. You'll incorporate a secret message into an LED transmitter that will be undetectable by the human eye. Then, you'll build a receiver that will decode the message!

The transmitter can be built easily on a breadboard without soldering; the receiver requires basic soldering of through-hole components.

A BRIEF HISTORY OF OPTICAL COMMUNICATIONS

There's a deep history of transmitting information over light waves both intentionally and accidentally. In what may have been the first electrical implementation, Alexander Graham Bell's **Photophone** from 1880 was able to transmit speech wirelessly by modulating sunlight reflecting off a mirror to a receiver several hundred meters away. The receiver would then convert the light back to sound. This invention pre-dated the commercialization of radio for wireless communication by decades. It wouldn't work on cloudy days when the sun was hidden, but still, Bell considered it one of his greatest inventions.

TIME REQUIRED: A Weekend

DIFFICULTY: Moderate

COST: $15–$65

MATERIALS

FOR THE DIY OPTICAL TRANSMITTER:

» **Arduino microcontroller board** I used an Uno R3 but any Arduino will work.

» **Red LED**

» **Resistor** Smaller values allow more current to pass through the LED and will result in a brighter output while sending data, which could increase your transmission distance, but pay attention to your LED's maximum current rating. Usually somewhere between 100Ω and 1kΩ will do the trick.

FOR THE RECEIVER:

» **OpticSpy Digital printed circuit board (PCB)** about $10 from oshpark.com/shared_projects/YRt9m2a8, or make your own. The Gerber files and complete BOM with DigiKey part numbers are available at my website grandideastudio.com/opticspy.

» **Fiber optic receiver module** Everlight PLR135/T9 or PLR237/T10BK

» **NPN transistor, 2N3904**

» **Inductor, 47µH**

» **Capacitors: 0.1µF ceramic (1) and 2.2µF tantalum (2)**

» **Resistors, 1kΩ (3)**

» **Voltage regulator, LDO, 3V, 100mA**

» **Green LED**

» **6-pin header**

» **Mini slide switch**

TOOLS

» **Soldering iron and solder**

» **Adafruit FTDI Friend (optional)** but very handy, $15, adafruit.com/product/284

JOE GRAND (Kingpin) is a computer engineer, hardware hacker, teacher, daddy, honorary doctor, and occasional YouTuber. He was a technical advisor during the founding of *Make:*, and this is his tenth article for the magazine.

Optical networks such as **VLC (Visual Light Communication)** and **Li-Fi (Light Fidelity)** are used as alternatives to traditional Wi-Fi wireless networks. Transmitters and receivers outfitted within a room or building will form the network. One advantage of these optical networks may be security — unauthorized access may be more difficult, as a hacker would need to intercept the light between the transmitter and receiver in order to monitor or inject data. A disadvantage is that the wavelengths of light used in these systems can't pass through objects (such as walls), so communication distance is limited.

OPTICAL SIDE CHANNEL ATTACKS

As for systems that unintentionally leak information over light waves, security researchers demonstrated in 2002 that many products, including network devices, hard drives, and printers, are susceptible to this type of *side channel attack.* Of the 39 devices they looked at, 14 were sending data through their status indicator LEDs that could be recovered and decoded by external receiver circuitry (toothycat.net/~sham/optical_tempest.pdf).

You'd think engineers would learn this lesson, but just recently researchers demonstrated (nassiben.com/video-based-crypta) that they could recover a private key (the secret piece of information used to encrypt and decrypt data in a secure system) from a device by capturing video footage of its power LED during a cryptographic operation! That's pretty scary and makes you wonder what other types of devices out there in the world that we rely on to be safe and secure are actually so.

Fiber-optic communications, which transmit data over modulated light through a physical medium of thin fiber rods made of glass or plastic, was developed in the 1950s and 60s. In the 1970s it was first used commercially for military and telephone applications. The technology is still used today, most recognizably as high-speed internet service to the home.

Ever play **laser tag**? It was originally used for military training with a system known as MILES (Multiple Integrated Laser Engagement System). It made its first appearance as a consumer product in 1979 as South Bend Toys' *Star Trek* Electronic Phaser Guns. These systems transmit and receive data over nonvisible light to simulate shooting or being hit.

CREATE YOUR OPTICAL COVERT CHANNEL

1. Build the transmitter

In order to create our own optical covert channel (sending and receiving data secretly through nontraditional, out-of-band methods), we need to build both a transmitter and a receiver. Building a transmitter to send data over light waves is super simple using modern platforms like Arduino or Raspberry Pi. Essentially, all we need to do is modulate the LED (turn it on or off) with our data at a rate faster than the human eye can detect. If done properly, the LED will appear to be continuously on even though it's blinking our data!

To accomplish this, we'll use a standard UART interface (also known as a serial port) to transmit the data with a **print()** or similar function. But, instead of printing text to a debug console, which is the most common use of UART, we'll redirect the data through an LED! Then, anyone with a receiver capable of capturing the light and decoding UART will be able to re-create the message.

For this transmitter circuit (Figures **A** and **B**), the LED will be on when the I/O pin RA1 is set to output **LOW** and the LED will be off when RA1 is set to output **HIGH**. Depending on the polarity of your UART interface (idle high vs. idle low), you might need to either change your UART configuration in software or "flip" the circuit so the LED's cathode is connected to GND and the high side of R2 (currently connected to VCC) is connected instead to the I/O pin. The goal here is to have the LED be solid on when data is not being transmitted and then blinking rapidly when bits are being sent.

You can download the complete Arduino code from my site at grandideastudio.com/opticspy, under Demonstrations; a snippet is shown in Figure **C**. You'll also find example code for the Parallax Propeller multicore microcontroller (parallax.com/propeller-1), the tiny Tomu board that fits inside a USB port (tomu.im/tomu.html), and even a TP-Link router!

2. Build the receiver

There are many ways to convert light into digital signals, most of which consist of a photodetector front-end and some amplification circuitry. I've created two open source receivers to make it easier to explore and experiment with optical data transmissions. Both can capture, amplify, and convert an optical signal into a digital form that can then be analyzed or decoded with a microcontroller, computer, logic analyzer, or other measurement tool. Complete build and usage details for both types of receivers are

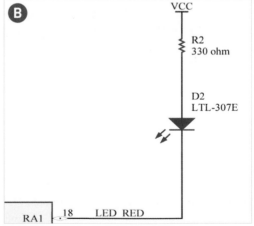

```
// Set up a new serial port
SoftwareSerial opticSerial = SoftwareSerial(rxPin, txPin);

opticSerial.print(msg_covert); // Transmit secret message through the LED
opticSerial.flush();           // Wait for all bytes to be transmitted
```

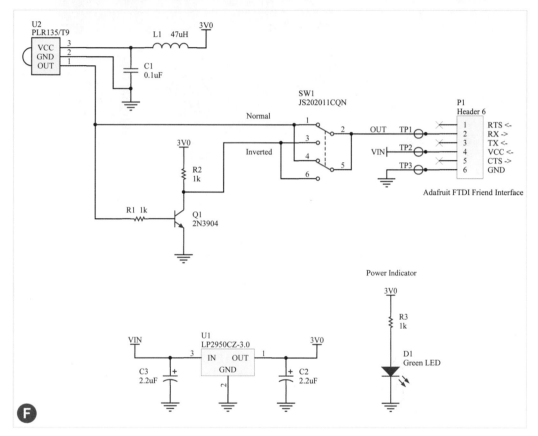

available at grandideastudio.com/opticspy.

For this project we'll use the **OpticSpy Digital receiver** (Figure **D**), based on an Everlight Photolink fiber-optic receiver module PLR135/T9 or PLR237/T10BK (Figure **E**). This module handles all optical interfacing and provides logic-level digital data output. It's commonly used for optical audio (TOSLINK) applications and contains all photodiode, amplification, and demodulation circuitry. When it receives light from a red LED with properly encoded data, it will output a digital version of the data on its OUT pin. The type of encoding scheme

supported by the module is known as **NRZ (non-return-to-zero)** — essentially a logic level **HIGH** corresponds to a **1** bit and a logic level **LOW** corresponds to a **0** bit, or vice versa), which is the same encoding used by UART interfaces! The LED needs to be placed all the way into the flap/door of the module in order for its light to be received by the photodiode.

For this receiver circuit (Figures **F** and **G** on the following page), U2 is the fiber-optic receiver module. C1 and L1 provide power supply filtering for the module. R1, R2, Q1, and SW1 provide an option to invert the data output by the module,

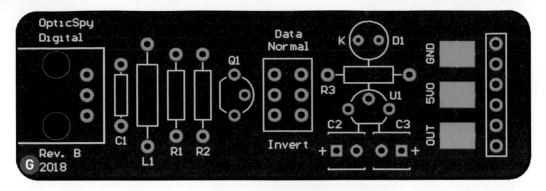

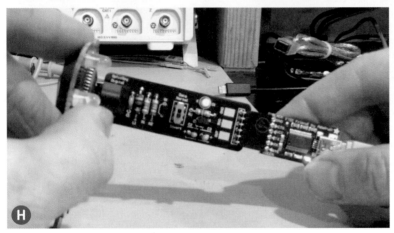

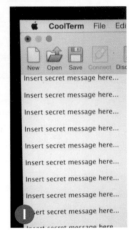

which may be required depending on the polarity of the transmitter's UART interface and the wiring of the transmitter's LED. U1, C2, and C3 comprise the voltage regulation, which will output a fixed 3.0V given an input of up to 30V. D1 and R3 serve as the power indicator where the LED D1 will turn on when power is properly applied to the system and the voltage regulator's output is active.

P1 is a 6-pin single row header to allow easy connection to power (VCC and GND) and the data output (RX). P1 is also compatible with Adafruit's FTDI Friend, which can plug into your host computer and provide power to the receiver while also providing a virtual COM port to decode and display the UART data directly on your computer using a terminal program like PuTTY, TeraTerm, CoolTerm, or Minicom.

3. Put it all together

With your transmitter and receiver assembled, you can finally send secret messages!

In Figure **H**, the receiver circuitry is connected to the FTDI Friend which is connected to my host computer running a terminal program. My LED transmitter is on the left, with the red LED pressed into the fiber-optic receiver module. The resulting message is displayed on the terminal program on the computer (Figure **I**). Congratulations, you have optically exfiltrated your secret!

FURTHER EXPLORATION IN PHOTONIC EXFILTRATION

This project is a basic implementation to get you started. If you're interested in having a longer-range optical communication system and/or the capability to capture and analyze modulation schemes other than NRZ, then consider building the more advanced **OpticSpy Analog receiver** on my website. With such a receiver you'll be able to capture optical signals from many more types of devices, such as status indicator LEDs from consumer electronics (Figures **J** and **K**) and iPhone proximity sensors.

You could also experiment with other types of transmitters, like different colors/wavelengths of LED, including nonvisible light such as infrared. Or try replacing the LED with the **Laser Diode Module Driver** on my site (Figure ), with a PCB from oshpark.com/shared_projects/WV8fBzyW.

Here are a few other projects that could supplement your toolkit:

- Forrest Mims' *Engineer's Mini Notebook: Optoelectronics Circuits*, originally published by Radio Shack in 1985, is a classic guide to all things optical. makezine.com/go/mims-opto
- Craig Heffner's IRis project is a very sensitive, high-gain amplifier designed to receive modulated IR signals from remote controls and proximity sensors on mobile phones. github.com/devttys0/IRis
- The Dark Art Lab shows how to convert a hobbyist audio amplifier kit into a laser microphone that can receive sound modulated by a vibrating surface. thedarkartlab.com/here-be-dragons/laser-microphone
- Chapter 5 of Michal Zalewski's book *Silence on the Wire: A Field Guide to Passive Reconnaissance and Indirect Attacks* covers optical exfiltration in great detail and has a design for an optical receiver that plugs into an old PC's parallel port.
- "xLED: Covert Data Exfiltration from Air-Gapped Networks via Router LEDs" is one of many projects from the Cyber Security Research Center at Ben-Gurion University of the Negev that look into all sorts of ways to exfiltrate data from air-gapped computers. arxiv.org/abs/1706.01140
- Ronen and Shamir's "Extended Functionality Attacks on IoT Devices: The Case of Smart Lights" demonstrates a practical attack using consumer IoT light bulbs for covert communication through varying light intensity levels. www.wisdom.weizmann.ac.il/~eyalro/EyalShamirLed.pdf

I hope your experiments are enlightening (get it?) and that you have fun with optical communications! ◉

RASPBERRY SPY STREAMER CAM

SURVEIL YOUR DOMAIN — AND STREAM IT TO TWITCH, YOUTUBE, OR A PRIVATE RTMP SERVER

Written and photographed
by Diego Gonzalez

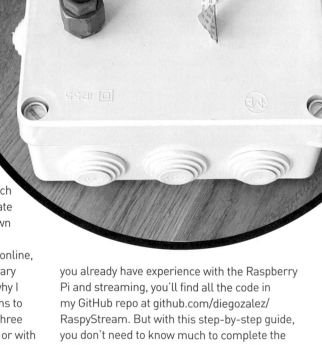

With this project, you'll turn a Raspberry Pi into a streaming spy camera that you can place anywhere with Wi-Fi or Ethernet, and it will upload its live video feed to an RTMP server. This format is commonly used for live streaming to Twitch or YouTube, but you can also make a private server and watch it or record it on your own computer.

There are a lot of projects like this one online, but most of them use the old *Raspivid* library which is not supported anymore. That's why I created this project, adapting old programs to the new *Libcam* library. It's composed of three shell scripts for streaming with no audio, or with music, or with a microphone. A fourth script runs your chosen stream in a screen session.

Building this project, you'll get more familiar with the Linux command line, and you can dig deeper into FFmpeg and its infinity of uses. If you already have experience with the Raspberry Pi and streaming, you'll find all the code in my GitHub repo at github.com/diegozalez/RaspyStream. But with this step-by-step guide, you don't need to know much to complete the project successfully, even as your first Pi project.

1. BUILD THE HARDWARE

First, you'll attach the camera to the box. To do this I use a flexible coolant pipe so I can position

TIME REQUIRED: A Weekend

DIFFICULTY: Easy

COST: $110–$120

MATERIALS

- » **Raspberry Pi 5 or 4B** with power supply. Older models will work but the new ones are recommended for their computing power.
- » **Raspberry Pi Camera Module 3** Mine is the wide-angle version and it comes with the cable.
- » **Pi case with heatsink (optional)** but recommended as it gets a bit hot
- » **SD card, 16GB**
- » **Electrical box** with rubber grommets. Mine is 15cm×11cm×8cm.
- » **Plastic flexible coolant pipe** such as eBay 283389046604
- » **Screw, M5** to mount camera to flex pipe
- » **Brass pipe fittings, ¼":** cap (1) and female to female (1)
- » **Screw, M4** with washers and wingnut
- » **USB microphone (optional)**

TOOLS

- » **Hand drill and bits**
- » **Computer and SD card reader**
- » **Box cutter or rotary cutter**
- » **Basic hand tools** like pliers and screwdrivers

the camera freely. I made a simple adapter out of steel plate, an M5 bolt, and four screws and standoffs (Figure **A**), but you can DIY something simple like gluing a camera case to the pipe.

The other end goes to a female-to-female ¼" brass pipe fitting. This connects to a brass cap that I drilled a hole in, to pass a screw that gets tightened through the plastic case lid with some washers and a wingnut (Figure **B**).

Now the camera's ribbon cable has to go through the box. I made a small slit with a box cutter and connected the camera cable to the Raspberry Pi and the camera (Figure **C**). Use the box's rubber grommets to pass the other cables.

2. SET UP THE RASPBERRY PI

To set up the Pi operating system on the SD card, I recommend using the official Raspberry Pi Imager, free from raspberrypi.com/software. Once that's installed on your computer, insert your SD card and select it in the configuration

DIEGO GONZALEZ is a freelance welder working in Barcelona, Spain. He likes to do electronic projects in his free time.

page (Figure **D**).

Click the gear icon for the Settings page, then select a hostname, enable SSH, set a password (Figure **E**). Also set up the Wi-Fi if you're going to connect it that way, but Ethernet cable is preferred. When you're done with settings, click on Write and wait for it to finish. Your SD card is ready.

Plug everything into your Pi: the SD card, the camera, USB microphone, Ethernet, and lastly the power supply (Figure **F**).

Now it's time to connect to the Pi from your computer. I use Putty (Figure **G**) but you can use any SSH tool. Under **host**, put **RaspyStream. local** and open; it will ask for the username and password. Use the one you entered into the Pi Imager before. The defaults are user **pi** and password **raspberry**.

Finally let's update everything so we don't have any problems later. Type the commands:

```
$ sudo apt update
$ sudo apt upgrade
```

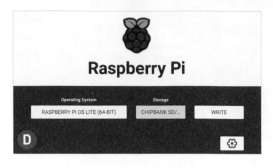

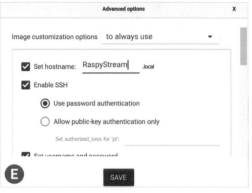

3. INSTALL PACKAGES

Now you'll install the necessary software packages so everything runs smoothly. First is FFmpeg, which is used to mix the audio and video and stream it to Twitch:

```
$ sudo apt install -y ffmpeg
```

To test the installation, type:

```
$ ffmpeg
```

And you should see something like this:

```
ffmpeg version 4.3.5-0+deb11u1+rpt3
Copyright (c) 2000-2022 the FFmpeg
developers
```

You also need Screen. This will help to run the streaming in the background. To install it, type:

```
sudo apt install -y screen
```

4. YOUR FIRST TWITCH STREAM

Now you can prepare everything to make your first stream. Since you'll be sending a lot of files between the computer and the Pi, I recommend using WinSCP (winscp.net/eng), a simple program that sends files over the internet.

Download the RaspyStream code repository

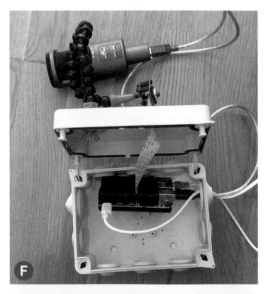

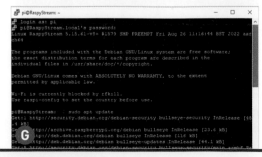

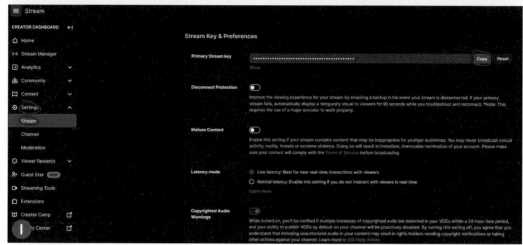

(github.com/diegozalez/RaspyStream) directly to the Pi and unzip it, with the following commands:
```
$ wget https://github.com/diegozalez/
RaspyStream/archive/refs/heads/main.zip
$ unzip main.zip
```

Go inside the folder *RaspyStream-main* by typing:
```
$ cd RaspyStream-main/
```

Here you will find the file *keys.txt*. Edit it using the Nano text editor:
```
$ nano keys.txt
```

Moving with the arrow keys, go to the line that begins **URL** (Figure **H**), and type your streaming URL; you can find this at help.twitch.tv/s/twitch-ingest-recommendation. To find your Twitch key, go to your Creator Dashboard → Settings → Primary Stream Key and copy it (Figure **I**).

Back in *keys.txt*, find the line that begins **Output #0** and add your Twitch key there. Now exit by pressing Ctrl+X and then Y to save the file.

For the *.sh* file to run, you need to make it executable:
```
$ chmod u+x ScreenIt.sh
```

And you're ready to stream! Just type:
```
$ ./ScreenIt.sh
```

After a few seconds, you'll see the console telling you the frame rate and other information. Go to your Twitch channel and see yourself LIVE!

5. TIPS AND TROUBLESHOOTING

That was fun but at some point, you need to stop the stream. Press Ctrl+C to stop it and the screen will terminate too. When you're running the stream you are using a Screen session, which means you can do other things while your stream is running. To exit the session press Ctrl+A+D, and you will now be able to see things like the Pi temperature with:
```
$ vcgencmd measure_temp
```

And check the Pi's resources with:
```
$ top
```

Exit with Ctrl+C. To go back to your Screen session type:
```
$ screen -r RaspyStream
```

If you have any problem, you can run the shell scripts directly, for example:
```
$ ./RaspyStreamMic.sh
```

This way you'll see the error message for troubleshooting.

6. TURN UP THE MUSIC

Now that you have a stream running, you might want to add music or a microphone. Let's start with the music. The script *RaspyStreamMusic.sh* plays *.wav* files from the music folder. To run it, first upload all your *.wav* files to the folder */Music* using WinSCP, then edit the file *Playlist.txt* with Nano and write the names of your files, just like the example in the file. Files will be played in the order you set here, and you can repeat a file if you wish. Once it finishes all, it will start again. Note that the default songs are blank.

Now just edit the script *ScreenIt.sh* with Nano:

```
$ nano ScreenIt.sh
```

Change where it says **./RaspyStreamMuted. sh** to **./RaspyStreamMusic.sh** and exit and save. You can now run:

```
$ ./ScreenIt.sh
```

And your stream, with music, will start like before.

7. ADD A MICROPHONE

You might want to stream microphone audio too. First connect your microphone and verify the Raspy code has detected it, with this command:

```
$ arecord -l
```

You will see something like this, showing your card number and device number:

```
pi@RaspyStream:~/RaspyStream-main $
arecord -l
**** List of CAPTURE Hardware Devices
****
card 3: U188 [U-188], device 0: USB
Audio [USB Audio]
  Subdevices: 1/1
  Subdevice #0: subdevice #0
```

Now edit the file *RaspyStreamMic.sh* with Nano:

```
$ nano RaspyStreamMic.sh
```

And go to the line where it says:

```
# Input_device is found by typing $
arecord -l
# example: card 2: U188 [U-188], device
0: USB Audio [USB Audio]
InputA=3,0
```

And modify the line **InputA=<your card number>,<your device number>** then exit and save.

Edit the *./ScreenIt.sh* script like in the music example, but this time with *./RaspyStreamMic.sh* then run it with the same command:

```
./ScreenIt.sh
```

That's it! You're live on the air!

8. LOCAL RTMP SERVER

This project is a great way to stream online, but sometimes you want to test the stream locally or set it up privately for secret spy stuff. For this we can use MonaServer, which is a small RTMP server, to stream from the Raspberry Pi to a computer of your choice. (Adobe's Real-Time Messaging Protocol was originally developed by Macromedia for streaming video to Flash Player, but today it's used for all kinds of video and audio streaming.) Then you can access the stream from that computer or others (Figure **J**), including your phone. To set up your RTMP server, follow these steps:

First, prepare Monaserver:
- Download Monaserver2 from github.com/ MonaSolutions/MonaServer2 for Windows, Mac, or Unix. I'm using Windows 64-bit in this example
- Extract the *MonaServer_Win64.zip* on your desktop
- Open the folder */MonaServer_Win64/*
- Right-click on *MonaServer.exe* and execute as administrator (Figure **K**).
- Allow access if Windows prompts a pop-up (Figure **L**).

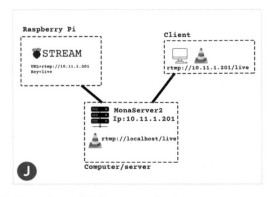

Next, change the server on the Raspberry Pi script:

- Open the script *RaspyStreamLocal.sh* and where it says **Ip** enter the computer's IP address where you set up MonaServer, for example
  `Ip=10.11.1.201`
- Then run the script:
  `$ ./RaspyStreamLocal.sh`

Finally, connect the client to the server:

- Download and install VLC (videolan.org/vlc) on the computer where you want to see the video feed. This can be the same as the MonaServer or other computers and phones with the VLC app in the same network.
- Open VLC and go to Media → Open network stream (Figure **M**).
- On the Network tab, write your MonaServer address as the network URL, for example
  `rtmp://10.11.1.201/live` (Figure **N**).
- And last, press Play.

Now you should see and hear your stream. The steps will be similar for Mac and Linux computers. You can also do this on a cellphone with the VLC app and watch your stream from anywhere!

LIVESTREAM IT!
You can use your Raspberry Spy Stream Cam in many situations, including security and surveillance, video podcasting, birdwatching or wildlife cams, or even watching plants in your garden. It's capable of 1080p resolution at 30 frames per second.

YouTube Live streaming works similarly to Twitch; learn more at youtube.com/howyoutubeworks/product-features/live.

IMPROVEMENTS
Now it's your turn to get creative!

- Make it a **night vision spy cam** by using a Pi NoIR camera and adding infrared LEDs.
- Make it a **portable movie streaming cam** by adding a power bank and a 4G modem to take it outside.
- **Add a light** for dark environments or video podcasting.

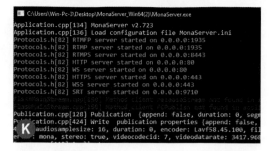

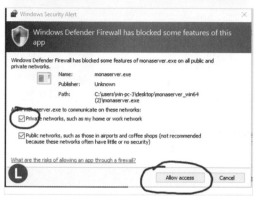

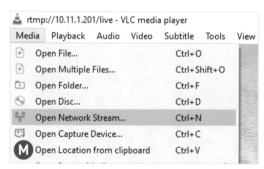

- Create your own **custom enclosure**, or add a ¼-20 nut to accept a standard **tripod screw**.

Let me know if you come up with more ideas. I would love to see if you make this project; you can find my contact info at github.com/diegozalez/RaspyStream or tag me on Instgram @abyteofthat. ◆

SEATTLE CRIME CAMS
2017

PHOTO: Christina Bakuchava

Seattle Crime Cams (2017) — City police share data on where they're going in real time. This project finds live cameras nearby, with live police radio.

THE ART OF SURVEILLANCE

PROVOCATIVE TECH ARTIST **DRIES DEPOORTER** EXPOSES THE MANY WAYS YOU'RE BEING WATCHED

Q&A with Keith Hammond

You're out sight-seeing and you post a selfie to Instagram. Before long, an unknown computer posts a video of you — taken before, during, and after your selfie! — from a nearby security camera. What the? Is Interpol after you? The NSA?

Actually it's just one guy. *The Follower* (2022) was the provocative work of Belgian technology artist **Dries Depoorter**, who exploits unsecured cameras and APIs to expose the dark sides of cybersecurity, privacy, social media, and artificial intelligence.

With a background in electronics, Depoorter also studied media arts and worked in advertising. Now he makes art for a living, and one of his main mediums is AI, applied to ubiquitous internet-connected surveillance cameras: security cams, traffic cams, doorbell cams, cameras in somebody's living room, lobby, or legislature.

I chatted with Depoorter to learn more about how he deploys AI to spy, and why.

How did you become focused on surveillance? Was there an incident that turned you in that direction?
Surveillance, privacy, AI, and the role of technology in society have always intrigued me. While there's no singular event in my life that steered me toward this topic, I've always been fascinated by the tension between the benefits of technology and the potential threats to our privacy.

How did you find open cameras for your work? Lists of public cams? Or did you hunt them down?
I mainly use Shodan, a search engine for internet-connected devices. Shodan provides an extensive database of cameras and other devices that are openly accessible on the internet. I haven't had the need to resort to wardriving.

What's your go-to technology for image recognition?
In most of my installations I use Raspberry Pi. If I need something more powerful I use Nvidia Jetson Nano; if I still need more power I will use AWS. In most of my work I use Keras for face recognition, and for object detection I use Yolo.

So what was the reaction to The Follower?
It was intense. The reaction was, "If only one guy achieved that with simple open source software, imagine what a government can do with access to way more cameras, way more computer power, and way more resources."

I had to take down the video because of copyright issues with public cameras, some companies claimed they had rights to the images.

In your view what are the most highly surveilled cities? I've heard London? Does the USA compare?
Yes, I think it's London. But in some countries it's easier to find open cameras.

Jaywalking shows how pervasive cameras are, but also how easily we turn into informants in a surveillance society. It reminded me of that psychology experiment where obedient subjects were willing to press a button to torture someone!
Absolutely. *Jaywalking* highlights not just the reach of surveillance but also how it nudges us to police each other. It's reminiscent of the Milgram experiment, where authority influenced people to

Dries Depoorter presenting in Copenhagen, 2018.

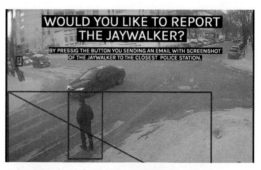

The Follower (2022) — Uses open cameras and AI to find out how an Instagram photo was taken. I recorded a selection of open cameras for weeks, then scraped all Instagram photos tagged with the same locations. Image recognition software compares the two.

Jaywalking (2016) uses live cameras at street intersections, some public, some just had the default password! I wrote image recognition software to detect when people are jaywalking. Then the screen asks viewers whether they want to report the jaywalker; if they press the big button, it emails the local police station with a photo of the jaywalker.

Quick Fix (2019–2023) — Installation allowing viewers to buy likes and followers. I currently have 50,000 fake accounts on Instagram and I can make sure they like or follow you. If I get asked by exhibitions I can produce more of these — I'm a huge fan of Raspberry Pi. No, the software is not open source!

Short Life (2020–2023) — A clock you can buy that shows how much percentage of your life is already done. I prototyped it with Arduino and then made my own PCBs. It uses [life expectancy] data from WHO for how old people get in every country. When you order it, you fill in your birthday and where you're from, and it's personalized; on the back I write your name. It's an "edition of 1 million" — we will only sell 1 million of it! **driesdepoorter.be/shop**

act against their morals. In both we see the power of external forces shaping individual behavior.

Quick Fix *points out problems of trust and fraud in social media: bots, fake accounts, inflated user statistics. But it also gets at the influencer economy specifically — if attention is money, then why not cheat to keep those numbers up?*
You've hit the nail on the head! *Quick Fix* shines a light on the hidden corners of social media. If I can manipulate numbers and attention, imagine what bigger entities with more resources can do.

It makes us question what's real and what's not, especially when popularity and money are so closely linked. It's a peek behind the curtain of the social media game, where sometimes, things aren't as they seem.

Your **Surveillance** *pieces give us the uneasy feeling that any camera might be connected to AI that is classifying and reporting what is sees — even hunting for someone specific. It's like the fantastical all-seeing NSA in the spy movies, but you're showing that potentially anyone can do it.*
Sure thing! These projects show us just how much surveillance in combination with AI is a part of our everyday life. It's clear that cameras and technology *are* watching in many places we go. It's a reminder for us to be aware and think about how much we're being seen.

With **The Flemish Scrollers** *are you intentionally turning the tables? Governments can use AI to spy on us to make sure we're not breaking the laws, but here we can use AI to keep an eye on the government while they're making the laws!*
It's interesting, right? *The Flemish Scrollers* plays with that idea, flipping the script on surveillance. It's about empowering the public and holding authorities accountable, reminding everyone that transparency should be a two-way street. ◕

KEITH HAMMOND is Editor-in-Chief of *Make:*.

Dries Depoorter, Boudewijn Bollmann, Bieke Depoorter / Magnum Photos

Surveillance Speaker (2018) — Tries to describe what the camera is seeing, with a really natural voice. One of the most beautiful moments I had with this piece was, there was a [set of] twins walking around the camera, and the piece was saying, "I see a girl taking a selfie in a mirror." So sometimes it's really poetical, and sometimes it's creepy accurate.

Surveillance Paparazzi (2020) — Microsoft has a paid API where you send in an image and it will recognize 200,000 celebrities around the world. So I made a bunch of computers that are constantly checking open cameras around Hollywood trying to find a celebrity. One screen shows the process, the other shows the results. They're all false positives. Includes doorbell cameras!

The Flemish Scrollers (2022) — Tagging politicians with AI and face recognition when they use their phones during government meetings livestreamed on YouTube. A Python script uses Keras software to search for phones and identify a distracted politician, then a video is then posted to Twitter and Instagram with the politician tagged, and the message: "Dear distracted <Politician Name>, pls stay focused!"

In galleries it's shown in a server rack that makes a lot of noise; I really like to show the hardware behind this kind of calculations. After I launched it, they changed the way they livestream — normally you had an overview of the whole room, but now they only film the speaker!

Border Birds (2022–2023) — I made this project with my sister Bieke Depoorter of Magnum Photos (**biekedepoorter.com**). I found a lot of open cameras on borders of two countries. I use machine learning to capture images of birds that are crossing those borders, then we print them out and sell them. **driesdepoorter.be/shop**

Boards Are Back

Written by David J. Groom

DAVID J. GROOM loves writing code you can touch. If he's not hacking on wearables, he's building a companion bot, growing his extensive collection of dev boards, or hacking on 90s DOS-based palmtops. Find him on Mastodon at @ishotjr@chaos.social

Mark Madeo

New evolutions in dev boards make this a metamorphic period for Makers.

In last year's Boards Guide, the overarching themes were scarcity and how vendors and makers coped with ongoing supply chain failures. Some industry experts even predicted that things might never fully return to normal, but thankfully the debacle is easing this year. As depicted on the cover, maker titans Arduino and Raspberry Pi are both back with new flagship boards in the form of the Uno R4 and Pi 5 respectively. As I type this, rpilocator.com is showing pagefuls of in-stock listings for pretty much the full Raspberry Pi product range, and I see hundreds of the latest Arduino Uno in stock at DigiKey, SparkFun, Pimoroni, and our own Maker Shed. To put it succinctly: boards are back!

Within this revivified ecosystem, we are seeing various exciting genera evolving. The alphas — or as I've been calling them, **beast boards** — such as DFRobot's LattePanda Sigma and Nvidia's Jetson Orin Nano, offer incredible computing power and performance, rivaling or even exceeding their desktop counterparts in some applications. Another interesting class of **all-in-one hybrids** has emerged from the primordial ooze, which feature built-in displays, sensors, and sometimes even cameras. Boards like the SparkFun Datalogger IoT, M5Stack CoreS3, and DFRobot Unihiker represent a complete project lab in one compact unit, allowing you to start diving into code right away, without any soldering or breadboarding.

Last year, we identified several **species of silicon** that were thriving in our challenging environment. RP2040, ESP32, and the new phylum of RISC-V-based chips continue to spread their DNA across the dev board ecosystem, even as other chips become available once again. Across this year's featured boards, we're still seeing a number of Espressif ESP32-based and Raspberry Pi RP2040 designs (build the PicoScope, an RP2040-based mini oscilloscope, on page 58), as well as the occasional STMicro,

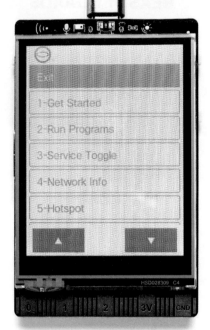

DFRobot Unihiker SBC with built-in display.

> North America is a bucket with holes in the bottom — we are pouring Pi 4s into Adafruit and Micro Center! —Eben Upton

Nordic, Realtek, GigaDevice, NXP, Broadcom, Renesas, and even Intel chip. While Arm still dominates, RISC-V chips appear to be climbing the silicon food chain. And if you want to create your own specimen, our **DIY silicon** piece on page 50 introduces you to services like Zero to ASIC Course and Tiny Tapeout that let you do just that!

The **Internet of Things (IoT)** biome has been evolving quietly, perhaps somewhat in the shadow of the burgeoning AI habitat, although the pace has picked up recently thanks to the increasing popularity and availability of LoRaWAN, enabling products like Seeed's SenseCAP T1000 LoRaWAN Tracker and Wio Tracker 1110. IoT behemoths Particle are back with a new Photon 2 Wi-Fi/BLE board (see "You're Muted," page 54) and a generous Free tier of their cloud platform that should far exceed the needs of most makers. And if you're really heading out in the wilderness, the

> Imagine Ford still makes the Model T, but it has evolved. That's Arduino Uno — it's a great tool for learning. —Massimo Banzi

SparkFun Artemis Global Tracker can send and receive short messages almost anywhere via the Iridium satellite network.

As boards evolve, so do the **language families** used to communicate with them. Adafruit has invested heavily in supporting their CircuitPython fork of the MicroPython compiler and interpreter over the past six years, and with support for many popular SAMD-, nRF-, and RP2040-based boards, plus a new web workflow in 8.x that allows ESP32-based boards to be used, the world of hardware development is opening up delightfully to a generation of Python software developers.

Meanwhile, Arduino have thrown their weight behind the original MicroPython platform, both in the form of their own "experimental" Arduino Lab for MicroPython IDE, and in their support of the OpenMV IDE. Raspberry Pi have also favored MicroPython with their Pico and Pico W boards and documentation, although developers can just as easily use CircuitPython if they prefer. UK-based serial innovators Pimoroni have also forked MicroPython to support their own RP2040-based products, as well as Raspberry Pi's Pico boards.

Despite Python's emergence as a great way to get started in embedded, we're still seeing C/C++ dominate overall, with most vendors taking a C-first approach to driver/library development and board support.

As species mature, they tend to evolve specialized adaptations to cope with specific environments. Companies like Adafruit, SparkFun, and Pimoroni have been cranking out **unique, special-purpose boards** targeting LED animation, cosplay, home automation, data logging, and other specific applications, often based around inexpensive RP2040 and ESP32 chips. Pete Warden's Useful Sensors has taken this concept even further, with devices that

Beast Wars

While single-board computers (SBCs) like Raspberry Pi and LattePanda may dwarf the power of smaller microcontroller boards, there are even mightier megafauna that focus on high-performance computing, typically for AI, industrial, and server applications. Compared to the quad-core Cortex-A72 processor found in the Raspberry Pi 4 for example, a Jetson Orin module contains six to twelve A78 cores. Combined with the Ampere GPU Tensor and CUDA cores, the six A78 cores in the Jetson Orin Nano Developer Kit provide 80x the AI performance of the previous entry-level Jetson Nano, with 6.6x the CPU performance — comparable to an Intel i7 on compute alone.

But when you're ready to unleash the ultimate beast, the Ampere Altra Dev Kit's 128 cores benchmark in the realm of 100 Raspberry Pi 4s, and with greater energy efficiency. These machines outdo even Intel's powerful Xeon D-2700 with several times the performance at a fraction of the energy cost, and the ability to add Nvidia GPUs for even more AI performance.

Alafia Ai 128-core Ampere medical imaging AI workstation with Ampere Altra Dev Kit with COM-HPC module by ADLINK Technology built-in.

Tomasz Swatowski / ADLINK, Camilo Buscaron / Alafia Ai, Adafruit, Seeed, Wilderness Labs

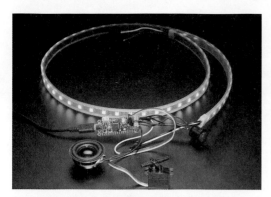

Adafruit's application-specific RP2040 Prop-Maker Feather, and Seeed's port-laden Grove Beginner Kit.

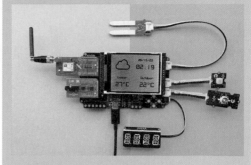

Over a decade ago, a curious mutation appeared on the maker landscape. In a world of C and C++, a new board emerged based on Microsoft's C# language and .NET framework. As was the style at the time, the board was named Netduino, for its Arduino Uno form factor. Its creator, Chris Walker, even wrote a Getting Started guide for *Make:* in 2012.

Fast-forward to 2023, and the Netduino, acquired by Wilderness Labs, has evolved into the Meadow project, with the Adafruit-style Meadow F7v2 Feather emerging from beta via the v1.0 Meadow.OS this year.

In the same way that Micro/CircuitPython enables software developers to dive into hardware development, the .NET Standard-compatible runtime allows an army of C# devs to create enterprise-level IoT solutions.

With an extensive API, fantastic documentation and examples, the available all-in-one Project Lab v3, plus copious detailed Hackster walk-throughs, Meadow is a great way to get started with embedded development for .NET devs — or anyone!

answer a single question like "Is that a person?" or "What does that QR code say?" Projects that once called for a general-purpose Uno or Feather, plus a stack of shields, breadboards, and breakouts, can now be accomplished with a single, purpose-built board like Pimoroni's Servo 2040 18-channel servo controller and Interstate 75 RGB LED matrix driver, Adafruit's Matrix Portal S3 CircuitPython-powered internet display enabler and RP2040 Prop-Maker Feather LED/audio/servo controller, and SparkFun's Thing Plus Matter MGM240P home automation nexus.

Excitingly, we're seeing more general-purpose boards that encourage **crossbreeding** thanks to the plethora of Stemma QT, Qwiic, and Grove ports that ease cross-vendor compatibility and obviate the need for breadboarding or soldering when just a few sensors or actuators are needed for a project. **Symbiotic relationships** between vendors have resulted in some interesting collaborations as well, such as the beefy new i.MX RT1010-powered M7 Metro from NXP, Adafruit, and DigiKey, and the exciting new XRP (Experiential Robotics Platform) kit from SparkFun, Raspberry Pi, DigiKey, DEKA, and others, which provides an inexpensive on-ramp

to FIRST Robotics with a free curriculum from Worcester Polytechnic Institute (see page 125).

The world of boards is flourishing again, thanks to renewed supply and increased diversity. It was incredibly challenging to narrow the side-by-side comparison in our 2024 Boards Guide, mailed with this issue, to just 81 standout boards, and harder still to pick 14 supreme specimens for our New and Notable list. If one theme has emerged across the ecosystem this year, it's that there's never been a better time to be part of this vibrant, ever-expanding community.

GARETH HALFACREE is a noted technology journalist based in Bradford, UK. He's the author of user guides for the Raspberry Pi, Pico, and BBC micro:bit, and an expert in educational and embedded computing.

Chip designs submitted through Matt Venn's programs are sent for production onto wafers like this one.

DIY Silicon!

Now you can design and manufacture your own custom chips — for cheap

Written by Gareth Halfacree

One of Venn's first chip designs turned a VGA monitor into a desk clock.

He learned that Edwards was maintaining an open-source **_digital synthesis flow_** — a complete tool chain for creating digital circuits, from the logic behavior source code to the finished physical layouts for fabrication — called **Qflow**. "So when I got home, I tried it out on one of my FPGA designs, and I got some GDS [electronic design automation] files out, and I was like, 'That's cool!'" Venn continues. "Then I thought, 'What about running a workshop where people could do this?'"

He looked into how much it cost to actually get a chip made — but the numbers were disappointing. "It's about $10,000 to do a short run. I just thought, 'It's not going to work, people aren't going to pay €1,000 [over $1,000] to take a course and get a chip.'"

But Venn wasn't satisfied just creating chip designs. "Just doing a course where you end up with the files is not interesting enough," he says. "I wanted to end up with a chip."

Going to Silicon

Then in 2020 Tim Ansell announced that Google would offer chip designers a **_tapeout_** — the lithography photomask for etching chips — to get their open source designs manufactured together on a shared silicon wafer (aka **_shuttle_**), for free. "I was, like, 'I'm in!' I was just totally ready," recalls Venn. "Luck favors the prepared."

That Google-funded program, known as **Open Multi-Project Wafer** or **OpenMPW** (developers. google.com/silicon), gave designers access to an open-source **_process design kit (PDK)_** for the 130nm manufacturing process at chip foundry SkyWater Technology. With a standardized shuttle layout for input and output, and access to the OpenLane toolchain on Efabless, suddenly chip production without major financial backing became possible.

What's a PDK? "It's got all the information you need to know to design a chip using a certain factory's process," Venn explains. "Having an open-source PDK where everything is public and you don't have to sign an NDA [non-disclosure agreement] is a game changer." In 2022 a second foundry, GlobalFoundries, opened their PDK for OpenMPW, giving makers access to a 180nm process as well.

Matt Venn is a man on a mission: to make it possible for any maker anywhere to design their own chips, going from the core concepts of how the layers in a semiconductor interact all the way up to making a fully functional, application-specific integrated circuit (ASIC) of their own — and receiving the physical chip in the mail.

"Even though I've been using microcontrollers for 20 years, I never thought that someone [like me] could make them," Venn recalls — this despite his history in working with field-programmable gate array (FPGA) chips, which allow you to customize their hardware internally. "It was so out of my idea of what was possible."

Open Source Chip Design

That is, until he attended an eye-opening Week of Open Source Hardware in Zürich, Switzerland. "Tim Edwards, who's the head of analog at Efabless, did a presentation on an open source chip where everything was open, apart from the RAM," Venn recalls. "He even had the board there running a little firmware printing out some messages onto an LCD, and it was using Claire Wolf's PicoRV32. That talk blew my mind."

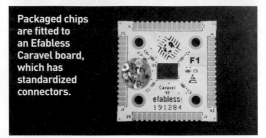

The tiny chips are produced at a SkyWater Technology fab using the company's open process design kit (PDK).

Packaged chips are fitted to an Efabless Caravel board, which has standardized connectors.

Some submitted designs are undeniably artistic, like this skull-shaped field-effect transistor.

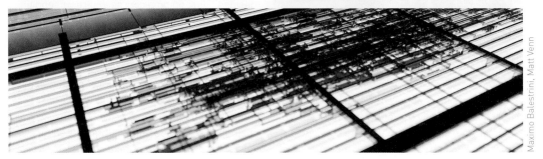

Maximo Balestrini, Matt Venn

Each Tiny Tapeout chip contains copies of every project submitted for that production run.

Zero to ASIC Course

SkyWater's open PDK was exactly the catalyst Venn needed to launch an educational course to bring chip design to a wider audience who, like Venn himself, had probably never considered that they could hold a self-designed chip in their hand.

"I did a talk about it for Hackaday Remoticon — this was in the pandemic," Venn explains. He asked participants if they'd pay $500 to learn how to design chips and get them made, and 200 signed up. "So I was like, 'Okay, I mean, let's go!' and then I spent one month solidly working and recording videos and writing up on my experience and launched the Zero to ASIC Course, and haven't looked back."

Backed by 6 hours of video content and 11 practical projects, **Zero to ASIC Course** (zerotoasiccourse.com) is designed to take attendees from a basic background in electronics to creating their own physical chip on an MPW — delivered already mounted on a PCB for testing. At $650, plus $50 for the test board, it's considerably cheaper than rival commercial training.

Tiny Tapeout

But Venn wanted to drop that barrier even lower with his next project, **Tiny Tapeout** (tinytapeout. com). Like Zero to ASIC Course, Tiny Tapeout is designed to introduce newcomers to chip design — but its lessons are free, with attendees only paying if they want the payoff of a physical chip. This starts at just $100, dropping the cost dramatically through a simple approach: putting multiple small designs on a single die to create a multi-project chip. Jumpers on the test board let you select your IC design from the dozens of others included in the chip — or to flit around the other submissions, if you want.

Nowadays anyone can blink an LED on a microcontroller by copying a simple script from the internet into Arduino's free IDE. "You can pretty much do that now for a custom chip, too," Venn claims. "With Tiny Tapeout you can go to **Wokwi**, which is this in-browser circuit design and simulation tool, connect an AND gate between two inputs and one output, and then go to our GitHub template system, run the ASIC tools as part of an automated action, and then submit it to the next Tiny Tapeout run — and six months later receive a board with a chip on it where if you press two buttons then the output light turns on.

"Of course, there's a whole world of stuff happening under the hood that is abstracted away, just like in the microcontroller example. You're benefiting from all the work that's been done to make it that easy. So, yes, it's now that easy for a random maker to make a chip."

For 100 bucks you get an IC about 100 by 160 microns, big enough to fit about 1,000 gates. What can you do with 1,000 gates? "Little blinkers, flashers, PWM servo drivers, motor drivers — people do whole CPUs." And you can pay more for more chip space.

Venn hopes Tiny Tapeout will become the de facto choice for small-scale chip production. "I'm hoping it's like the OSH Park of open-source silicon. Then if you want to learn Verilog, or you want to learn formal verification, or you want to end up doing a whole chip design, then you upgrade to the Zero to ASIC Course."

SiliWiz

There's a final string to Venn's bow, and it's there to help you get your head around what exactly is happening at the very lowest of levels. **SiliWiz** (app.siliwiz.com) is an online interactive tool to learn how semiconductors work and how you build them using lithography. "We have visibility now all the way down to the electrons, basically, so we've got an unprecedented level of transparency on how chips really work," Venn says. "And I think that's one of the exciting things: you don't need to know how it all works, but you could do if you wanted to." ●

The projects on each Tiny Tapeout chip are wired such that you can jump between yours and the others at will.

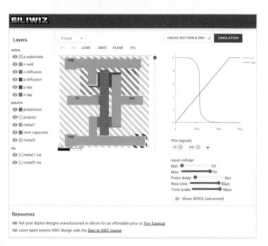

"Yes, it's now that easy for a random maker to make a chip."

The SiliWiz software runs in-browser and offers SPICE-powered circuit simulation.

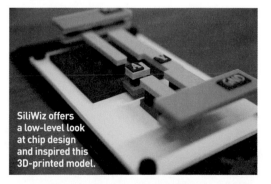

SiliWiz offers a low-level look at chip design and inspired this 3D-printed model.

'You're Muted!'

Oops! Use machine learning to automagically unmute your mic when anyone utters the immortal phrase **Written by the Particle Team**

PARTICLE

This article is the result of the collaborative work of many folks in our global company who love hacking on hardware and software! particle.io

TIME REQUIRED: 2 Hours

DIFFICULTY: Easy/Intermediate

COST: $25–$30

MATERIALS
» **Particle Photon 2 microcontroller** $18 from store.particle.io/products/photon-2. You can buy the following parts separately, or get them all plus lots more sensors and LEDs in the Edge ML Kit for Photon 2, particle.io/products/edge-ai-kit-for-photon-2.
» **PDM digital microphone module** with header soldered, Adafruit 3492
» **Breadboard** any size that fits both modules
» **Jumper wires, male-male (6)**
» **USB cable** for connecting Photon to your PC

TOOLS
» **Computer with Chrome web browser**

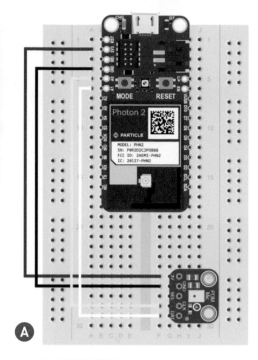

(A)

o you occasionally forget to unmute when speaking at a conference or during calls? Embarrassing!

This little detector tool uses machine learning to recognize the spoken phrase "You're muted" — and then send the keystroke to your computer to automatically unmute you!

The You're Muted project is powered by Particle's brand new Photon 2, a next-generation Wi-Fi development board with a powerful enough processor to handle *edge ML* — machine learning on an edge device, like an embedded computer or microcontroller. A powerful speech recognition model is coded in TensorFlow Lite, trained online using Edge Impulse, and then deployed to your Photon 2 microcontroller where it runs locally without needing any cloud computing or even internet access! You're Muted celebrates the leveraging of modern MCUs, silicon, and ML to bring voice command experiences, akin to Siri and Alexa, into the microcontroller domain.

Enjoy this simple, fun, easy-to-rig tutorial. The only software required is a web browser to get going. For the more adventurous, you can use our development environment, available as a Visual Studio Code extension, to customize to your heart's content. We'll walk you through everything you need, from wiring up the project, flashing firmware to your Particle device, and configuring your laptop to respond to the Photon 2's commands.

1. LET'S GET WIRING!
Simply place your Photon 2 and PDM digital microphone module on the breadboard as shown in Figure **Ⓐ**, and connect them together with four jumper wires as shown. The Photon 2 contains a 200MHz CPU and 4.5MB of SRAM, enough to run the machine learning (ML) workload of identifying words in an audio stream.

PDM digital microphones are different from the I²S digital mics you may have used before. They use a digital audio scheme called *pulse density modulation* that most 32-bit processors can handle. You can use a variety of PDM microphones with this project, but we selected the Adafruit PDM mic as it's readily available and has a good noise floor to make it able to pick up your voice from across the room.

The pinout for this project is as follows:

Mic module	Wire color	Photon 2	Details
3V	Red	3V3	3.3V power
GND	Black	GND	Ground
SEL	—	—	Left / right select. Typically this goes unconnected.
CLK	Yellow	A0	PDM Clock
DAT	White	A1	PDM Data

Particle Team

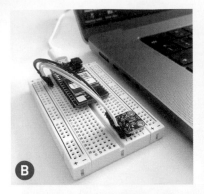

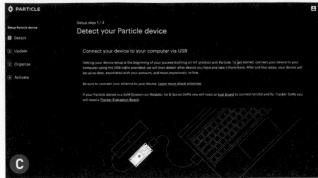

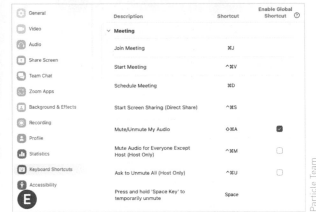

2. PROGRAM THE PHOTON 2

Plug the Photon 2 module into your Linux, Mac, or Windows computer (Figure **B**). Boot up the Chrome web browser on your computer and navigate over to particle.io/makemagazine2023. This page will walk you through programming the Photon 2 from within your browser environment, connecting over WebUSB (Figure **C**).

> **NOTE:** You'll be prompted to set up a Particle account if you want to benefit from the platform features (such as over-the-air programming), but it's not required for flashing your Photon 2.

After flashing, the Photon 2 might prompt you to select the Keyboard Type (as it emulates a keyboard that can send keystrokes to your computer!). Select ANSI if prompted (Figure **D**).

Once programming is complete (about 2 minutes), the Photon 2 will be running the entire "You're Muted" example application including the local ML processing engine. It should work out of the box with Zoom calls — after you test it you can customize it for other apps and your own hacking!

3. CONFIGURE YOUR CONFERENCE CALL APP

We're about to unleash the magic! These instructions are tailored for Zoom, but fear not, most conference apps support keyboard shortcuts in a similar way in order to trigger the Mute function and other actions on your computer.

- **For Zoom:** Go to Settings → Keyboard Shortcuts (Figure **E**) and find Mute/Unmute My Audio. Check the box for Enable Global Shortcut, and jot down the shortcut. On a Mac, it's Command-Shift-A, and on Windows, it's Alt-A. You can also modify these shortcut keys if you want to remap them, but you'll need to update the Photon firmware as well to match!

- **For Google Meet and Microsoft Teams:** No modifications are needed to these applications as they use fixed keyboard shortcuts, but you must recompile the Photon 2 firmware in order to change the keys that the Photon 2 is sending over USB to trigger these applications.

Conference Call Nirvana

Ready for the grand test? Let's try the "You're muted" detector inside your conference app! Go ahead and open up a Zoom call.

Your Photon 2 will boot up with the following LED behaviors:

- **Green** = Listening on the microphone
- **Orange** = Detected phrase and sent a keypress to the computer
- **Red** = Error of some kind — check your wiring please!

Now utter the immortal phrase "You're muted" and check that the Photon 2 recognizes your words and triggers the keypress to your computer.

HOW IT WORKS

So what's going on in this demo? A TinyML-based model is running on the Photon 2, part of an application that you can download and rebuild in the GitHub repository (github.com/particle-iot/make-magazine-muted-demo). Audio samples are captured and listened to continuously from the microphone and these are fed into the ML model. This model returns a confidence factor for each set of audio samples as to whether it contains the words "You're muted" and if there is high enough confidence, a keypress is sent to the computer using the built-in mouse and keyboard library inside the Photon 2.

This TinyML model was trained on roughly 5 minutes of different people saying "You're muted." While that's roughly the order of samples you need to make a viable keyword detector, it doesn't mean this sample set is perfect, and not all variations of accents will work out of the box. We've included instructions in our GitHub repo on how to add in your own voice samples via a web application and then regenerate the model to tune it to your local accent or even change the words being used to trigger off completely!

CUSTOMIZING YOUR DETECTOR

Building the firmware for your device enables a world of exploration and modifications. For a detailed tutorial on how to reprogram your Photon 2 and get started with the Particle Developer Experience, please visit particle.io/makemagazine2023.

```
void sendUnMuteKeys()
{
    Log.info("Sending unmute key!");

    //Select which key to send to the computer!

    // MAC OS:
    // Keyboard.click(KEY_A, MOD_LEFT_COMMAND | MOD_LSHIFT);

    // Windows:
    Keyboard.click(KEY_A, MOD_LALT);

    // Linux:
    // Keyboard.click(KEY_A, MOD_LEFT_CONTROL | MOD_LALT);
```

A high-level view of the steps involved are:

- Download the GitHub repository to your laptop
- Install Visual Studio Code and the Particle plugin
- Open the project folder by going to *File → Open Folder*
- Compile your project either locally (don't worry, we'll download the toolchains for you!) or using our cloud compiler (it's free!)
- Flash your Photon 2 over USB with the rebuilt project and confirm everything still works
- Modify the keypresses that the device sends — you can find the easy-to-modify code at the top of *make-magazine-muted-demo.cpp* (Figure **F**) in the project folder!

To replace the keyword and enable your Photon 2 to trigger on other phrases ("You're on mute," "Unmute yourself," "Hey dummy," etc.) or to simply add new training samples to the existing data set, please see the detailed instructions in the *README* file in our GitHub repo.

WHAT ELSE WILL YOU AUTOMATE?

Now that you've got your feet wet in the world of edge ML and the Particle device ecosystem, it's time to start automating everything else in your life. We've always wanted a toothbrush timer that automatically starts when we start brushing our teeth. How about you?

Happy tinkering, and may your Photon 2 always have your back while you're on Zoom! ●

• • • This project was developed by Nick Lambourne, Alberto Sanchez, Rick Kaseguma, Marek Poliks, Kent Mok, and Mohit Bhoite. • • •

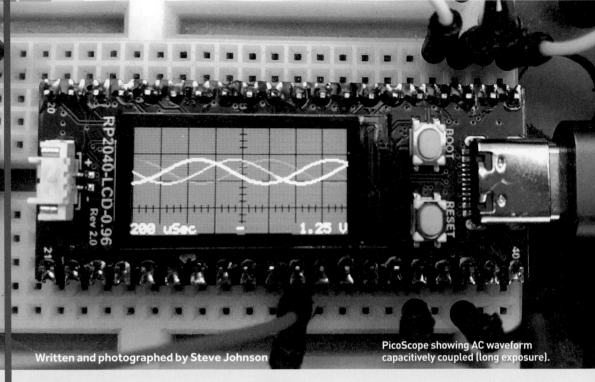

PicoScope showing AC waveform capacitively coupled (long exposure).

Written and photographed by Steve Johnson

PicoScope

Build a useful mini oscilloscope for your workbench, for about 25 bucks!

When I was in high school, I had a horrible DuMont oscilloscope that had probably last seen service in World War II. It did help me with some of my beginner projects, though. Later, as an undergrad and early in my career, I had access to Tektronix and Hewlett-Packard scopes, which spoiled me with their stability, features, and usability. More recently, I've picked up a surplus Tek oscilloscope, but it's pretty big and bulky. I've often needed an oscilloscope at my desk, but there's no room for it. I decided to create my own, using the Raspberry Pi Pico!

Back to BASIC

Despite professionally programming in C for most of my career, Hewlett-Packard BASIC was my first "high-level" programming language. I've used BASIC on HP desktop calculators, minicomputers, time-sharing systems, Commodore 64s, Z80-based CP/M, Microsoft Visual BASIC, and pretty much anything else you can imagine.

I've also been playing with the fabulous MMBasic from Peter Mather and Geoff Graham, on the ESP32 and on Windows. In the last year or so, Peter has ported MMBasic to the Raspberry Pi Pico, a $4 microcontroller board that may be more familiar as a target for Arduino and Python programming. The Pico version is called PicoMite. The MMBasic community, largely based in the UK and Australia / New Zealand, has a rich, welcoming, and very active developer community

TIME REQUIRED: `1-2 Hours`

DIFFICULTY: `Easy`

COST: `$20-25`

MATERIALS
» **Waveshare RP2040 microcontroller board with LCD display** with pre-soldered pin headers, RP2040-LCD-0.96, $14 from waveshare.com
» **Solderless breadboard** Adafruit 5422 or equivalent, $5
» **Miniature SPST pushbutton switches (3)** about $1 each
» **Breadboard jumper wires, solid, 18–20 gauge**
» **USB-C cable**

TOOLS
» **Wire stripper**
» **Needlenose pliers**
» **Soldering iron and solder (optional)**
» **Computer with free software:**
  • **PicoMite firmware** geoffg.net/picomite.html
  • **Terminal emulator** such as TeraTerm, tera-term.en.lo4d.com
  • **PicoScope software** github.com/sfjohnso/PicoW/blob/main/PicoScope.bas

Steve Johnson is a software developer, musician, and science fiction fan in Southern California.

on The Back Shed Forum (thebackshed.com/forum). Variants of the interpreter are available for the Pico and the Pico-W, the latter with extensions for TCP/IP and other internet protocols.

Here's how to make your own mini o-scope inspired by classic designs and programmed in PicoMite MMBasic on an RP2040 microcontroller.

Build Your Mini Oscilloscope
1. PREPARE HARDWARE

To make my life easier, I've picked up display boards from Waveshare, a Chinese manufacturer. The one I use for this project is the Waveshare RP2040-LCD-0.96, a Raspberry Pi Pico-style board that includes an RP2040 chip, an 80×160 full-color LCD display, and uncommitted I/O pins carried out to the same castellated edge-connectors and through-holes as on the original Pico. This board costs about $14 with pre-soldered pin headers, and has a USB-C connector that provides power and communication. Board documentation is available at waveshare.com/wiki/RP2040-LCD-0.96.

To use the board "bare bones," you can clip a signal lead to the scope's input castellated pin, and a ground lead to a nearby GND pin. If you're more ambitious, you can solder the headers, or buy the version with pre-soldered headers, and use it with the prototyping board of your choice.

This setup works great for signals ranging from 0VDC to 3.3VDC, the reference voltage for the Pico's analog-to-digital converters (ADCs). If you anticipate needing to view low-voltage AC signals (like the audio output from an MP3 player, for instance), you'll need a few other components that you may already have around the workshop. The photo on the opposite page shows a 1kHz waveform capacitively connected to the Pico.

The PicoScope program can be used with either a terminal emulator or with three pushbuttons to configure sweep rate, vertical scale, and triggering methods. If you'd like to use the PicoScope without a dedicated connection to your computer, you'll also need three momentary-contact SPST pushbutton switches.

2. INSTALL MMBASIC FIRMWARE

The PicoMite firmware is available free from Geoff Graham's website, at geoffg.net/picomite.html. Download the firmware to your computer, and unzip into a directory of your choice. The package comes with an extensive manual with plenty of programming examples, periodically updated to include new features that arise through Peter Mather's excellent work. The firmware is in a file with a *.uf2* extension.

Connect the Pico to the cable, and while holding down the boot button on the Pico, plug the cable into an available USB port on your computer. When you release the button, the Pico should appear as another drive. Copy the UF2 file onto this new drive. Once the copy operation is complete, the Pico will automatically reboot,

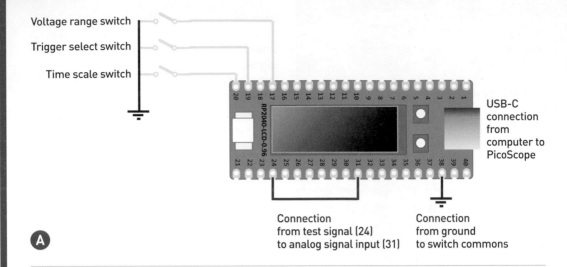

Voltage range switch

Trigger select switch

Time scale switch

USB-C connection from computer to PicoScope

A

Connection from test signal (24) to analog signal input (31)

Connection from ground to switch commons

and the new drive will disappear, and a new USB-based serial port will appear in Device Manager. Make a note of it there, or just play with the terminal emulator (next) until you get the connection working.

You'll need a terminal emulator to communicate with the Pico board. The PicoMite documentation suggests using TeraTerm on Windows, which is freely available at tera-term.en.lo4d.com. Once you've downloaded and installed it, start it up, and start a Serial communications session (8 bits, 1 stop bit, no flow control, any baud rate) using the new USB Serial device as the target. Press the Enter key a couple of times; the PicoMite firmware will automatically recognize the baud rate and respond with a command prompt (>).

3. CONFIGURE THE PICO LCD BOARD
The PicoMite firmware can interact with lots of different hardware, but it needs to be told what you have. Here's how to do that. (More information about what this all means can be found in the PicoMite manual, under Options, around page 79.)

First, type **OPTION LIST** at the > prompt. It should respond with the firmware version, and another prompt. (The **OPTION LIST** command is extremely useful for debugging; you'll probably use this again and again.)

Next, type in these commands to configure the board:

```
OPTION SYSTEM SPI GP10,GP11,GP28
OPTION COLORCODE ON
OPTION POWER PWM ON
OPTION HEARTBEAT OFF
```

B

```
OPTION CPUSPEED 250000
OPTION DISPLAY 50, 132
OPTION LCDPANEL ST7735S,
LANDSCAPE,GP8,GP12,GP9,GP25
```

4. WIRE UP THE BOARD

All of the above can be done without doing any wiring! In order to use the PicoScope, you'll need to connect the analog input to either the test signal, or to some other 0–3.3V signal source (Figure **A**).

Gently insert the RP2040-LCD-0.96 into the breadboard. If you're using the nice Adafruit Solderless Breadboard for Raspberry Pi Pico, be sure to align pins 1 and 40 with the appropriate rows so that the pin labels are correct.

To see the test signal, insert a wire between pins 24 and 31. Either use a plug wire from your junk box, or strip 5mm of insulation off each end of an 18 to 20 gauge solid-core wire, and insert into the appropriate rows on your prototyping board (red wire in Figure **B**).

If you'd like to use switches instead of the terminal emulator to change voltage, timing, or trigger, wire up three SPST switches as shown in Figure A. These are the yellow and black wires in Figure B.

5. EDITING ON THE PICOMITE

The PicoMite firmware comes with a built-in editor that supports syntax coloring, and provides a very rapid turnaround for software development. Entering and leaving the editor are single-keystroke operations from the command prompt. Two of the **OPTION** commands (above) configured the editor to use syntax coloring and to display a wide and tall screen on your computer (Figure **C**).

If you prefer to use a different editor on your computer (I like Notepad++), it may be more comfortable for you to edit your programs there, then transfer the program file to the Pico. The firmware supports the venerable XMODEM protocol, again with single keystrokes at the Pico command line. However, TeraTerm needs to be configured to point to the directory on your computer in which you've saved your programs.

To transfer a program to the Pico, press F11 in the terminal emulator. The Pico will respond

with **XMODEM Receive**, telling you that it's ready to receive an XMODEM transmission from your computer. Then, click [**F**]ile, [**T**]ransfer, [**X**]modem, [**S**]end, and select the file to be transferred. TeraTerm will send it to the Pico, which will respond with the number of bytes transferred, and a command prompt. Enter the Pico's editor, and verify that all the lines of the program have been transferred.

The PicoMite firmware uses flash memory on the Pico for both a small virtual drive, and a work area to store programs, options you've set, and other information (Figure **D**). As you work with programs on the Pico, you can save them to either or both. I generally save to one of the four flash "slots" or the small drive every half hour or so. Once I'm settled on a program version, I also save it to my GitHub repository.

6. THE PICOSCOPE PROGRAM

Download the PicoScope program listing from my GitHub repository, github.com/sfjohnso/PicoW/blob/main/PicoScope.bas.

My PicoScope program makes use of structured program techniques, with subroutines

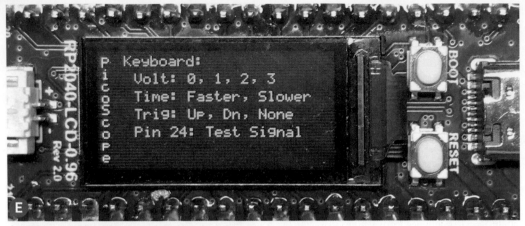

Waveshare board with the PicoScope title page displayed.

and functions for major operations. These are called by an ***initialization*** step, and a ***main loop*** (this structure should be very familiar to Arduino programmers). The program also makes extensive use of the graphics and built-in math functions that so enrich the BASIC programming environment provided by MMBasic through the PicoMite firmware.

The program also makes use of hardware-driven ***interrupt service routines***. These cause temporary interruption in the main processing loop to take care of short-duration real-world events, in this case, momentary-contact push-button presses to change vertical and horizontal scales and scope triggering.

The program includes a set of comments that identify itself and give credit to the members of The Back Shed Forum who, over the years, have developed some tips and tricks that I use.

Next, you'll see **OPTION EXPLICIT**. This makes sure that only declared variables can be used in the program; it also helps debugging.

In the program listing, you'll see a long list of constants. These are set up so that the program can't inadvertently change them. The use of constants helps debugging, too.

Next, you'll see a list of "dimensioned" variables and arrays. These hold values that do change throughout the execution of the program. Some of the arrays are loaded by the Pico's ADC subsystem to acquire the signal to be displayed, and others are initialized to be used later for the really, really fast graphics routines provided by the PicoMite firmware.

Again, to aid in debugging, you will see occasional use of **LOCAL** variables with functions and subroutines. These are only visible within those routines, and can't be seen or modified by other code.

The initialization portion of the program calls the following subroutines:

- **Initialize_Arrays:** Loads the arrays with values that allow the use of fast graphics routines, and sets up the various possible values for sweep rate (time), triggering, and voltage scales.
- **Initialize_Hardware:** Configures I/O pins on the Pico, sets up a pulse width modulated (PWM) test signal on a Pico pin, and creates a framebuffer and layer (in memory) for flicker-free graphics.
- **Display_Instructions:** Displays the keystroke commands that can be used on the terminal emulator to control the oscilloscope (Figure **E**).
- **Draw_Graticules:** Uses MMBasic's fast graphics commands and the initialized arrays to draw the oscilloscope gridlines, and time, trigger, and voltage indicators.

The main loop runs forever (or until the program is interrupted with a [Ctrl]-[C] keypress). During the main loop, switch presses trigger the interrupt service routines.

The first subroutine call is to **Randomize_Test_Signal**. This inserts some jitter in the test signal. The scope is so stable that you might not believe that it is running, unless you see a little bit of jitter!

The next statement clears a built-in timer that watches for no trigger. If after 2.5 seconds, no trigger is seen, a **No Trigger** message is displayed.

Another nested loop is opened:

This loop runs until a trigger condition is found in the acquired waveform. While waiting, the loop calls **Handle_Keypresses**, which looks for commands that the user may have entered on the keyboard via the terminal emulator, and **Handle_Switches**, which looks for user input from the front panel switches. These cycle through the timing ranges, the trigger types, and the voltage ranges.

The code then calls **Get_Samples**, which acquires enough signal values to span two screen widths, and **find_trigger**, which looks for a trigger condition in the first half of the samples. This guarantees that a screen width of signal can be displayed after the trigger is found. If a trigger is found, the loop exits. If not, additional samples are acquired. However, if the 2.5 second timer expires, the **Display_No Trigger** subroutine is called.

Eventually, a trigger is found, or the user changes the trigger condition to **Free Running**. When this happens, the code checks to see if it needs to clear the **No Trigger** message, and redraws the graticules if so, by calling the **Draw_Graticules** subroutine again.

The input voltage range is 0 to 3.3V. This needs to be scaled to the selected voltage scale by calling the **Scale_Samples** subroutine.

The scaled signal is then painted on the screen using MMBasic's fast graphics firmware by calling the **Update_Display** subroutine.

Use Your PicoScope

That's it! You've got a working mini oscilloscope that fits on any workbench and lets you visualize — actually see — just about any kind of electrical signal that varies over time: frequency, amplitude, noise, and so on (Figures **F**, **G**, **H**, and **I**). (You can learn more about using an oscilloscope at learn.sparkfun.com/tutorials/how-to-use-an-oscilloscope.)

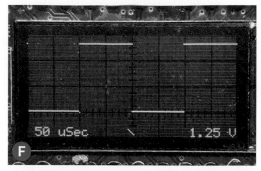

50 microsecond (µs)/div, negative (\) trigger, 1.25 volts/div test signal.

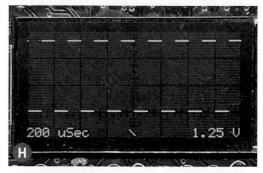

200µs/div, negative (\) trigger, 1.25 volts/div test signal.

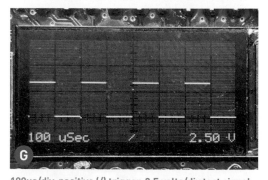

100µs/div, positive (/) trigger, 2.5 volts/div test signal.

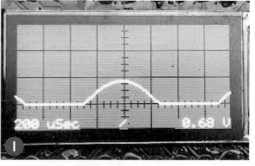

AC waveform not capacitively coupled.

Fast Math and Display Routines

The PicoScope also lets you show off your new MMBasic skills on the Raspberry Pi Pico. One of the many things that appeal to me about the MMBasic language and its PicoMite implementation is the careful attention paid to developer needs by Geoff Graham and Peter Mather. Much of this appears in extensions to graphics and math routines.

Two main subroutines in my code make use of the array extensions to the **LINE** and **PIXEL** graphics routines, which normally take single pairs of coordinates. The extensions will take arrays instead, automatically drawing dozens to hundreds of pixels or line segments in a single command. Take a look at the **Display_Graticules** and **Update_Display** subroutines for examples.

My **Initialize_Arrays** and **Scale_Samples** subroutine makes use of the extensions to normal BASIC math functions in MMBasic. Those extensions provide ways to rapidly pre-set arrays to a specified value, or obtain the min, max, mean, median, or many other derived values from arrays in single statements.

Box It Up

The PicoScope code, together with the Waveshare RP2040-LCD-0.96, would make a dandy battery-powered hand-held oscilloscope with the appropriate packaging. 3D-printed or laser-cut fabrication techniques would both work nicely. My neighbor Luis Mateo and I collaborated on the design of a simple enclosure that he created using his laser cutter (Figures **J** and **K**).

Make It Two-Channel

As I was writing this article, participants on The Back Shed Forum challenged each other to implement a two-channel oscilloscope. I was able to modify the PicoScope code to acquire and display a second channel without changing the program structure. The new code is also on my GitHub as *TwoTrace.bas*. It uses both pins 31 and 32 (GP26 and GP27) for 0–3.3V inputs. Because the RP2040's analog-to-digital converter multiplexes between the two inputs, the maximum display speed is cut in half, to 100µS per division (Figure **L**).

I hope you find this project useful, both to become aware of and make use of the PicoMite version of MMBasic, and also to implement a useful desktop miniature oscilloscope! **⊘**

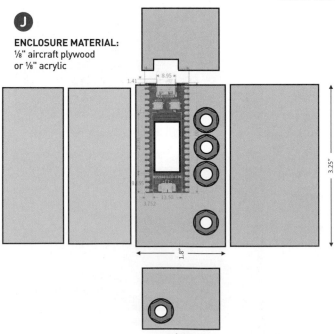

J

ENCLOSURE MATERIAL:
⅛" aircraft plywood
or ⅛" acrylic

Acknowledgments:

- *MMBasic (implemented for the RP2040 as PicoMite firmware) by Peter Mather and Geoff Graham*
- *Many ideas for this program were shared on The Back Shed Forum*
- *Sampling, Test Signal, Triggering, and Framebuffers from stanleyella*
- *Math Routines and Memory Copy from Peter Mather*
- *Luis Mateo for collaboration on the enclosure design*

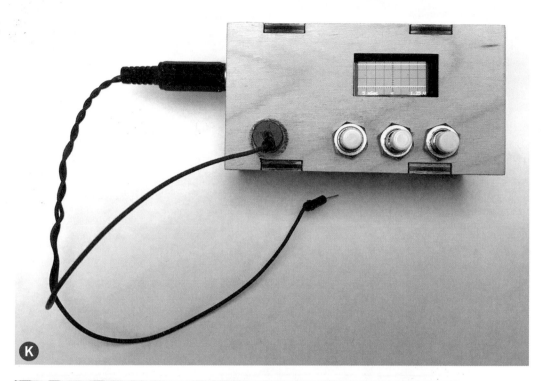

A two-channel version of the software works, too.

Bag to the Future

Build an illuminated, animated, tessellated tote using LED pebble lights and 3D-printed fabric

Written and photographed by Debra Ansell

DEBRA ANSELL is a maker and educator who will never stop demonstrating that LEDs improve everything.

I'm thrilled that this illuminated tote bag is my seventh (!) project write-up to appear in *Make:* magazine. Coincidentally (or perhaps not), my very first *Make:* article from 2017 (makezine. com/projects/led-matrix-handbag) described an LED matrix handbag that displayed animations and scrolling text. This new geometric bag is both retrospective and forward-thinking, as it retains its predecessor's mix of playfulness and practicality while incorporating more recent maker methods.

Since 2017, novel creative techniques, like 3D printing on fabric, and newly available products, such as uber-flexible LED strings, have continued to expand the potential of wearable tech. My new Tessellated Tote integrates a broad array of maker skills including 3D printing, sewing, and electronics to produce a modern carryall. The bag's front panel is illuminated by lightweight addressable string lights secured behind translucent tiles which are 3D printed onto a layer of tulle fabric. The tulle holds the individual tiles in a space-filling geometric arrangement that moves and flexes with the fabric around it.

In addition, the tote holds a hidden homage to scientific discovery in its front panel. The 3D-printed diamond/triangle tessellation pattern represents a newly discovered type of exotic condensed matter — a *bosonic crystal* composed of particles called *excitons* packed tightly into two overlapping atomic lattices with this tessellated pattern (space.com/exotic-new-state-of-matter-discovered-from-ultradense-crystal). When turned at a slight angle to one another, the lattices interact in a moiré pattern that traps the excitons just so.

Of course, the Tessellated Tote is fundamentally a fashion accessory. You don't need to recognize its futuristic tech or catch its subtle scientific references to enjoy its attractive design and colorful LED animations. And you can make it quite inexpensively depending on your choice of fabrics and LED controller.

Whichever aspect of this project most appeals to you, be it futurism or fashion, the result is a pretty and practical satchel that puts the "fun" back in "functional."

TIME REQUIRED: A Weekend
DIFFICULTY: Intermediate
COST: $30–$90

MATERIALS

» **3D printer filament, translucent PETG** such as Amazon B07ZNG4L9P
» **Tulle fabric, white, 12"×15"**
» **Faux leather or felt, 12"×14" or bigger** or other stiff, non-fraying fabric that can be cut with a vinyl cutter or laser cutter, for the tote's front panel. Glowforge Leatherette, Amazon B0B1GV7B14, cuts well with either.
» **Heavyweight fabric, ½ yard** for tote bag exterior. I used my favorite snakeskin-patterned faux leather, Joann's 18242560.
» **Lining fabric, ½ yard** Upholstery weight, or thinner cotton with an ironed-on fusible backing. Get more if you'll add pockets.
» **Zipper, at least 15" long** nylon coil zipper or zipper tape, Amazon B0BW8RTZL4
» **Webbing, 1" wide, 24" lengths (2)** or longer, for straps, Amazon B09B6W122Y
» **Thread** to match exterior and lining
» **Flexible LED "pebble" light string, 5V, 50mm spacing, 88-pixel length** addressable WS2811 RGB, such as AliExpress 3256805578356358
» **Cable, USB-A to 5.5×2.1mm male barrel jack** Amazon B07519D95L
» **LED strip controller, Pixelblaze V3** from shop.electromage.com; or your choice
» **Portable phone charger power bank, 5V, with USB-A port** at least 1A current, such as Amazon B09Z6R5XZ7
» **Hot glue sticks, translucent or clear**
» **Masking or painter's tape, ½" or narrower**
» **Male/female JST connector (optional)** for LED string, if not already attached
» **Connector, 5.5×2.1mm female barrel jack to wire terminals (optional)** if not already attached, Amazon B07795WSGP

TOOLS

» **Sewing machine** with both standard and zipper foot
» **Fabric scissors**
» **3D printer**
» **Small neodymium magnets or Kapton tape** to secure tulle mesh to the print bed
» **Hot glue gun**
» **Ruler**
» **Sewing clips or straight pins**
» **Small screwdriver** for screw terminals
» **Wire cutter or flush cutters**

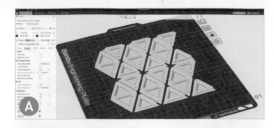

1. PREPARE THE PRINT FILES FOR THE FABRIC

Download the design files from the project page at makezine.com/go/tessellated-tote. The full lattice of 3D-printed tiles measures about 11"×13", larger than most desktop printers' print beds. The grid can be printed in four (or more) smaller sections. If you'd prefer to subdivide the grid yourself, the full design file is in *GeometricTilesAll. stl*. Otherwise, two subsections of the grid are in the files *GeometricTilesPartial1.stl* and *GeometricTilesPartial2.stl*. Print two copies of each to generate the entire front panel.

Import the STL file into your slicer program to see the arrangement of triangular and diamond tiles. Each tile is widest at its base and top, with a narrower "neck" in the middle. This narrow section secures each printed tile into a hole in the bag's front fabric panel. The tiles' back faces are recessed to hold the LEDs.

Print the grid of tiles with the flat face down for a smooth finish (Figure A). Set infill density to 100% and layer height to 0.2mm or finer. If your slicer supports variable layer heights, printing thinner layers near the tile's flat front face will reduce overhang issues where the tiles are filleted. Inside the slicer, position the tile grid in the middle of your print bed so there will be space to secure the tulle around it on all sides.

Set the slicer to pause the print at close to

2.6mm layer height. The cross-section of each tile is narrowest here, and this is where we will insert the tulle fabric into the print. Cut a piece of tulle that is smaller than your print bed but larger than the grid section you're printing, and set it aside for the next step.

2. PRINT THE TRANSLUCENT TILES

Durable PETG is the best choice of filament for the thin-edged tiles as it will withstand everyday wear better than PLA. Select a clear or translucent filament color. Insert the filament into your printer and start the print. When the printer pauses at 2.6mm, lay the cut piece of tulle over the printed tile layers, and use the small neodymium magnets or Kapton tape to secure it to the print bed around the tile grid perimeter (Figure B). Pull gently around the edges of the tulle to eliminate wrinkles, so that it lies flat over the existing print.

Restart the print. The small holes in the tulle mesh allow the next layers of filament to bond with the bottom layers, sandwiching the fabric in between. When the print is done, remove the tiles from the bed carefully so as not to rip the tulle. You now have a flexible mesh of 3D-printed shapes (Figure C). Repeat this procedure to print all sections of the grid. While the printing is taking place, you may proceed with the handbag construction in the following steps.

3. CUT THE TOTE FABRICS

Cut two 15"×17" rectangles from the lining fabric.

Cut one 15"×17" back piece, two 2½"×15" strips and two 2½"×12" strips from the main fabric.

Because the front tile panel contains many geometric holes, it's easiest to cut this piece with a laser cutter or vinyl cutter. Use the vector file *TilePanel.svg* to cut the front panel from your leatherette or non-fraying fabric.

> **WARNING!** Never laser-cut fabric containing vinyl/PVC or you can generate toxic chlorine gas.

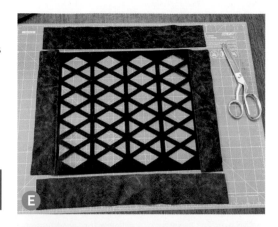

4. SEW BORDERS TO TILE PANEL

Align the 2½"×12" fabric strips with the short edges of the tile panel, right sides together, and secure with clips or pins (Figure **D**). Sew together, then fold the strips open with the seam underneath the strip. Topstitch the strips so they lie flat (Figure **E**).

> **NOTE:** All seams are ⅜" unless otherwise indicated.

Next, take the 2½"×17" strips and align them, right sides together with the long sides of the tile panel assembly. Sew the pieces together, then fold the strips open with the seam allowance underneath the strip. Topstitch along the fabric strips with the same technique as before.

The tile panel piece will form the front side of the bag (Figure **F**). It may now be slightly larger than the other 15"×17" bag panels. If so, trim its edges symmetrically to make it the same size as the other panels.

If you wish to add pockets to the bag lining (Figure **G**), now is the time. There's not enough space to describe the process here, but many online tutorials explain how to sew patch pockets (blog.treasurie.com/sewing-pockets-how-to-sew-a-pocket) or zipper pockets (so-sew-easy.com/add-zipper-pocket-purse-pattern) into a bag's interior.

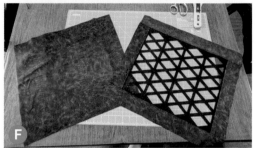

Place the lining pieces and exterior fabric pieces together. Cut 1½"×1½" squares out of the bottom corners of the both the lining (Figure **H**) and main fabric "sandwiches." These cutouts will form the bags' corners.

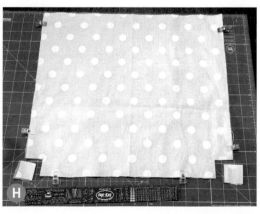

5. SEW THE BAG EXTERIOR AND STRAPS

Pin the two exterior fabric panels with right sides together. Sew straight lines to join both sides and the bottom of the exterior fabric pieces (red lines in Figure ), but do not sew the corner cutouts.

Then open the bag, pinching the unsewn corners together, opening the seam allowance, and folding the cutout edges together so that the seams align (Figure ). Sew directly across the folded cutout, going back and forth a few times for reinforcement to create the "boxed" corners (Figure K).

Cut two 24" lengths of webbing (or whatever length straps you prefer) and sew them to the right side of the bag body, equidistant from the center on both sides of the bag, with the strap ends aligned with the bag opening (Figure L). Stitch the straps in place ¼" below the bag opening.

6. SEW THE BAG LINING

The Tessellated Tote incorporates a zipper into the lining bottom to allow easy access to the electronics. Pin the zipper to the bottom edge of one of the bag lining pieces (Figure M), with right sides of zipper and lining together. The zipper ends should overhang the cutouts on both sides of the lining.

Using a zipper foot, sew the zipper to the lining, then fold the zipper away from the lining with seam allowance underneath the lining. Topstitch along the lining (Figure ). Next, place the unsewn zipper edge along the remaining lining piece, right sides together, and sew together with a zipper foot. Open up the zipper edge, fold the seam allowance under the lining and topstitch along the lining (Figure O).

Place the two lining pieces, which are now connected by a zipper along their bottom edges, right side together and pin them along their sides. Sew a straight seam down each side, following the red lines in Figure P.

Open the lining, pinching the square corner cutouts together so that the ends of the zipper stick out. Pin the corner sides together so that the middle of the zipper aligns with the side seams. Make sure that the zipper pull is located between the lining's two corners before you sew

O

P

them closed (Figure Q).

Pin the corner seams together and sew across the corners, using the same technique you did with the bag exterior. After sewing the corners, trim off any extra part of the zipper that hangs outside the lining (Figure R). The lining is now taking shape with boxed corners and a zip-open bottom (Figure S).

Q

7. JOIN INTERIOR AND EXTERIOR

Turn the main bag piece right side out and the lining wrong side out. With the lining zipper unzipped, insert the bag body into the lining (Figure T), right sides together, aligning the seams on both body and lining. Pin or clip the body and lining openings together, and sew around the circumference of the opening with a ½" seam allowance (Figure U).

Pull the bag exterior through the zipper to invert the bag so it's right side out (Figure V). Next, push the interior into the exterior, then smooth out and pin the main bag opening around the seam (Figure W). Topstitch around the bag opening to make it lie flat.

8. INSTALL THE TILES

Invert the bag again so that it is inside out and pull the lining away from the body. Gather

R

S

T

U

V

W

First LED →

the tulle-connected tile grids, trim the excess tulle from around the outside edges of the grid segments, and gently push the tiles, flat side first, into the holes in the front bag panel until the narrowest part of each tile is seated in the fabric (Figure X). While maneuvering the tiles into place, carefully stretch the panel fabric around the tiles to avoid pulling hard on the tulle.

Once all tiles are in place, turn the bag right side out through the lining and use a fingernail or butter knife to settle the fabric into the narrow channel in each tile. When finished, the tiles will sit flush with each other, just above the bag surface (Figure Y).

9. GLUE THE LED STRING

Turn the bag inside out again. Cut a length of LED string containing 88 LEDs and solder a JST connector to its input end. Connect it to your controller and power it up to make sure all the LEDs are working. Notice that, when illuminated, each LED displays a brighter front side and dimmer backside.

The LED string attaches to the tile grid in a vertical zigzag pattern along the tile columns. Each LED nestles, bright side downward, into the center of a tile's depression. Note that each of the diamond-shaped tiles contains two depressions and holds two LEDs in adjacent columns.

With the backside of the tile grid facing you, place the first LED in the string into the lowest leftmost triangle and run the LED string upwards along the first column of triangles as shown in Figure Z. Using skinny strips of masking or painters' tape, secure the string in place as you lay the rest of the string into alternating up and down columns. Place the tape *between* the tiles and away from the triangular depressions in the tile's centers (Figure Aa), to keep it out of the way of the hot glue in the next step.

Next, plug in the hot glue gun, and allow it to heat up. The hot glue serves a dual purpose by holding the LEDs in place and diffusing their light throughout each tile. Starting with the first LED in the string, place the nozzle of the glue gun in the center of the tile, just underneath the LED, and slowly squeeze a layer of glue about 5mm deep to cover the bottom of the tile's indentation (Figure Bb). Gently press the LED bright side down into

Bb

Cc

the glue, leaving it secured in place while the glue sets. Repeat this process for all 88 LEDs.

Once all LEDs are secured and the glue has cooled, remove all strips of tape, and turn the bag right side out through the lining. The bag construction is now complete (Figure Cc) and it's time to light things up!

Dd

10. SET UP THE PIXELBLAZE CONTROLLER

I use the Pixelblaze controller for many different projects because its software makes it easy to use the same LED animations on any arrangement of pixels. Section 4 of the Pixelblaze Pillows build in *Make:* Volume 83 describes the steps to attach a Pixelblaze to an LED string and connect to the software in a web browser (makezine.com/go/pixelblaze-pillows). Follow the same sequence of steps to join the LED string and 5V power supply to the controller (Figure Dd), then connect to the Pixelblaze software over Wi-Fi.

Once connected, open the Mapper tab in the Pixelblaze software, and cut and paste the contents of the downloaded *TesselatedToteMap.txt* file into the editor window (Figure Ee). Press the Save button below the editor to set the new LED layout.

There are several options on the Settings tab that can help to keep the power draw low and extend battery life. Most importantly, cap the LED brightness at or below 50%. Additionally, it is possible to run the Pixelblaze CPU at lower

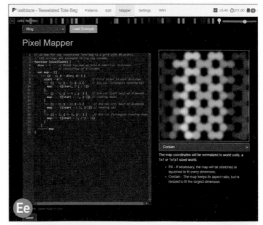

Ee

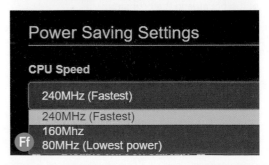

Ff

Gg

Hh

speeds than the default 240MHz; selecting the 80MHz option helps reduce energy consumption (Figure Ff). And if you won't need to connect to your Tesselated Tote's controller while wearing it, check the box to disable Wi-Fi on startup.

The best way to showcase your tote bag's tessellating geometric display is to curate a selection of LED animation patterns designed to look good in two dimensions (Figure Gg). Using the Playlist feature on the Patterns tab (Figure Hh), you can compile a set of patterns from the default list, as well as from patterns you write yourself in the Pixelblaze's JavaScript-based editor. You can set a default time for each animation to play, and you can advance through the list by physically pressing the button on the Pixelblaze controller.

THIS BUILD IS IN THE BAG

Much of the fun of creating this project came from experimenting with different tessellating tile patterns [see "Skill Builder: Tiles and Tessellations," on page 112]. You're free to create different geometric designs if you keep the spacing between your 3D-printed tiles the same as, or slightly smaller than your LED spacing. Fortunately, the flexible LED string used in this

project is available with a variety of LED intervals, ranging from 1.5cm to 10cm.

You can also adapt the techniques in this project to create different styles of illuminated, tessellated bags, such as a backpack or sling bag. Any bag pattern containing a flat, unadorned front panel and a lining that can be modified to allow access to the bag interior will work. Simply insert an opening, zippered or otherwise, into the lining design, and before sewing the bag exterior, cut holes corresponding to your 3D-printed tile layout into its front panel.

If you enjoy coding your own LED patterns, there are an endless variety of ways to customize the Tessellated Tote's illuminated display. You can also make it interactive: connecting a Pixelblaze Sensor board to the controller allows it to respond to sound and motion with dynamic animations.

The Tesselated Tote's combination of 3D printing, sewing, and electronics provided a great way to stretch my maker skillset while building on the original 2017 design. As I continue to learn new techniques from others in the maker community, I'm excited to see what creative ideas will work their way into further sequels to this "Bag to the Future!" ✪

Written and photographed by Doug Stowe

Ginormous Froebel Blocks

Super-size a classic teaching toy and put the *play* back in *playground*

DOUG STOWE is a woodworker, teacher, and writer in Eureka Springs, Arkansas, where he established the Wisdom of the Hands woodworking program at Clear Spring School. He's the author of 100-plus magazine articles and 14 books; his most recent, *The Wisdom of Our Hands: Crafting, a Life*, proposes a renewal of American education through hands-on learning.

In 2016 Blue Hills Press published my book *Making Classic Toys That Teach,* about the wonderful teaching tools developed by Friedrich Froebel (Fröbel), the inventor of kindergarten in the mid-1800s. The geometric toys that Froebel called *gifts* — including cubes, spheres, planks, cylinders, and prisms — were of enormous influence on the world's culture, having directly shaped such notables as architects Frank Lloyd Wright and Buckminster Fuller, toymaker Milton Bradley, and artists Paul Klee, Wassily Kandinsky, and Josef Albers. Froebel's gifts also influenced and predated the development of the Montessori educational method.

As the son of a kindergarten teacher, I realized the importance of play as an instrument in education. My mother would explain to parents at school conferences that when they would ask their children what they'd done in school today and their answer was "Play," that was the right answer.

At one time, under the inspiration of Froebel's kindergarten, educators wondered how to extend instructive play into the upper grades, and manual arts education was introduced in Scandinavian schools as a result. Scandinavian manual arts training, also known as educational *sloyd,* became important throughout the world for an all too short period of time.

I think history is important for two reasons. One is to avoid the failures of the past. The other is to provide a path for renewal of those things we've forgotten. Kindergarten is one of those things that ought to be remembered and replicated. If we were to each look seriously at the times in which our own learning has been energized, we would design schools to engage the power of play in all subjects and at all levels.

My book, published after years of study, attempted to empower parents, grandparents, and fellow woodworkers to do what generations had done before: make the gifts of early childhood that once served as a pivot point for an international revolution in education, one that has been largely abandoned despite its obvious value. Frank Lloyd Wright, in his autobiography, recalled his years of playing with Froebel's gifts: "These were smooth maple-wood blocks. All are in my fingers to this day."

A reproduction set of "Froebel gifts."

Friedrich Froebel (1782–1852), educator, advocate of play-based learning, inventor of kindergarten.

SUPER SIZE IT

In 2018 I began making super-sized Froebel blocks and placing them unceremoniously on the Clear Spring School playground. I knew of no one else doing such things and was curious what the student response would be. Even with just two or three blocks, before a full set was in place, the students knew immediately what to do with them: stack and arrange, just as they would do with smaller blocks. The complete set that Froebel had called Gift 3 consisted of eight cube blocks (though of much smaller size) that could be used to construct a larger cube, or be arranged as

A

Froebel advised into "forms of beauty," "forms of life" (building representative forms), or "forms of knowledge" (developing counting and stacking skills).

With the first eight giant blocks completed and seeing students' enthusiastic response, I began making Gift 4 at the same scale. This set consists of eight flat "tile" blocks that also can also be arranged into a large cube.

The students at Clear Spring School keep the blocks in a state of continual rearrangement. One day they'll be arranged as a fortress with space inside. The next they may be an obstacle course with kids taking turns to jump from one block to another.

Here's how to build super-sized Gifts 3 and 4 for your playground, using just a few tools and basic carpentry skills. They're made from ½" exterior plywood with some interior bracing to withstand rugged play and being left outdoors throughout the four seasons (Figure A).

TIME REQUIRED: 1 Hour per Block

DIFFICULTY: Easy/Moderate

COST: $50 per Block

MATERIALS / CUT LIST

For the Cube blocks (Gift 3):
- » **Exterior treated plywood, ½" thick, 4'×8' sheets** You'll need 6 sheets to make all 8 blocks, or ¾ sheet per block, cut as follows:
  - • **2 pcs. 24"×24"**
  - • **4 pcs. 23½"×23"**
- » **Exterior treated deck boards, 1¼" (5/4") thick, 6"×8'** You'll need 3 to make all 8 blocks. Rip them into 1¼" widths, then you'll need 24 linear ft. per block, cut as follows:
  - • **4 pcs. 1¼"×1¼"×23"**
  - • **8 pcs. 1¼"×1¼"×20½"**
- » **Deck screws, #8×1-5/8" long** You'll need 480 total or 60 per block.
- » **Construction adhesive**

For the Tile blocks (Gift 4):
- » **Exterior treated plywood, ½" thick, 4'×8' sheets** You'll need 8 sheets to make all 8 blocks, or 1 sheet per block, cut as follows:
  - • **2 pcs. 24"×48"**
  - • **2 pcs. 11"×48"**
  - • **3 pcs. 11"×23"**
- » **Exterior treated deck boards, 1¼" (5/4") thick, 6"×8'** You'll need 4 to make all 8 blocks. Rip them into 1¼" widths, then you'll need 32 linear ft. per block, cut as follows:
  - • **4 pcs. 1¼"×1¼"×11"**
  - • **6 pcs. 1¼"×1¼"×20¼"**
  - • **8 pcs. 1¼"×1¼"×21¾"**
- » **Deck screws, #8×1 5/8" long** You'll need 480 total or 60 per block.
- » **Construction adhesive**

TOOLS
- » **Table saw**
- » **Cordless drill with 5/64" or 1/16" bit** to pilot #8 screws
- » **Screwdriver bits** for deck screws
- » **Impact driver (optional)**
- » **Router with ½" radius round-over bit**
- » **Sandpaper**
- » **Long clamps (optional)** but helpful for glue-up

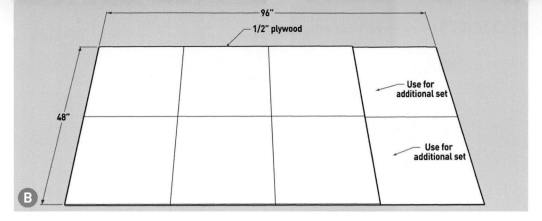

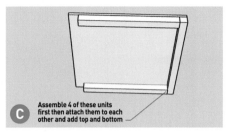

Assemble 4 of these units first then attach them to each other and add top and bottom

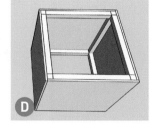

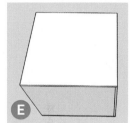

MAKING CUBES (GIFT 3)

Cubes are the simplest shape, measuring 2'×2'×2'.

1. CUT

Cut the plywood to size as listed in the cut list and shown in Figure **B**. Then rip the ⅝" (1¼"-thick) treated deck boards into 1¼" wide strips, and cut those to lengths described in the cut list.

2. ATTACH STRIPS

Use construction adhesive and screws to attach strips to three edges of the box sides, leaving one edge to be connected to the next side (Figure **C**).

3. ASSEMBLE

Use adhesive and screws to assemble the four sides to each other (Figure **D**), then to attach the top and bottom (Figure **E**).

MAKING FLAT TILE BLOCKS (GIFT 4)

The tiles, 1'×2'×4', require additional reinforcing in the middle, so require a slight difference in approach.

1. CUT

Begin by cutting the plywood to size according to the cut list and Figure **F**.

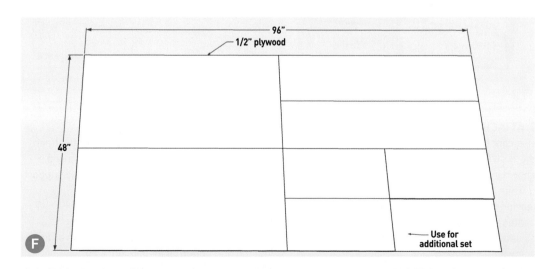

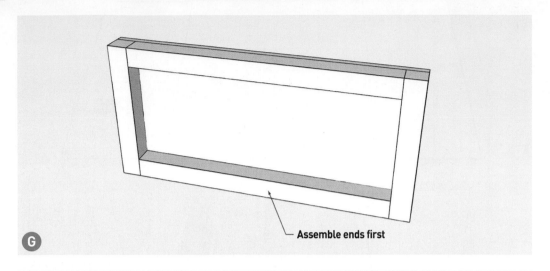

Assemble ends first

G

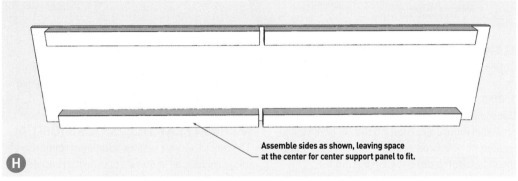

Assemble sides as shown, leaving space
at the center for center support panel to fit.

H

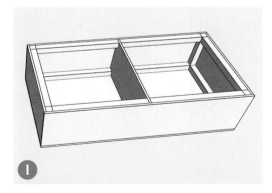

I

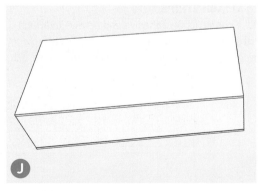

J

2. ATTACH STRIPS

Use construction adhesive and screws to attach the ⁵⁄₄ strips to the edges of the end panels, as shown in Figure **G**.

Then attach the strips to the side panels, leaving space for the internal support at the center and at the ends to lap over the end panels, as shown in Figure **H**, again using adhesive and screws.

3. ADD REINFORCING PANEL

Use construction adhesive and screws to attach the sides to the ends. Next, use adhesive and screws to attach ⁵⁄₄ blocking to the center support panel, and screw it in place (Figure **I**).

4. ASSEMBLE

Use construction adhesive and screws to attach the top and bottom panels (Figure **J**).

FINISHING

For each finished block, use a router with a ½"
radius round-over bit to smooth the edges.

Finally, sanding the corners is also advised.

PLAY TIME!

Just like the tiny kindergarten blocks, your eight
giant playground cubes can be arranged into
one humongous cube, and so can the eight giant
tiles. To freshen ours up and protect against the
elements, we recently sealed them with some
recycled house stain (Figure **K**).

The advantage of the Froebel blocks over a
conventional playground is that the continual
redesign by the kids themselves keeps it fresh.
The students use the blocks in the traditional
ways — making representative shapes and
"forms of beauty" — and make playgrounds of
their own design. Their engineering skills and
cooperative play skills all come into play. You can
see the blocks in action at youtu.be/Si-ukJoeDwY.

ADDING MORE GIFTS

Over the years other instruments of play have
been added to our Froebel playground: wooden
planks, plastic 55-gallon barrels (cylinders), and a
cut-in-two tractor tire (arcs). ◗

K

LEARNING MORE

- *Making Classic Toys That Teach: Step-by-
  Step Instructions for Building Froebel's
  Iconic Developmental Toys,* Blue Hills
  Press, bluehillspress.com/shop/making-
  classic-toys-that-teach

- **Froebel USA:** youtube.com/@FroebelUSA

- **Froebel's influence on art, design,
  architecture:** 99percentinvisible.org/
  episode/froebels-gifts

Join the Air Movement!

Build your own "Corsi box" air purifiers to battle viruses, wildfire smoke, and indoor air pollution

Written by Christina Cole and Victoria Jaqua

Removing infectious aerosols and wildfire smoke particles from indoor air has been a DIY, community-led project since August 2020, when Dr. Richard Corsi and Jim Rosenthal designed the Corsi-Rosenthal Cube. Now colloquially known as a *Corsi box*, these DIY air purifiers have been adapted by indoor air quality (IAQ) experts and validated as an effective air cleaning method by several peer-reviewed studies. They're easy to make and they work better than some commercial units!

The robust performance of Corsi boxes is largely due to the electrostatic charge (which attracts particles) and large surface area (pleated/accordioned into a compact size) of ordinary MERV-13 filters you can buy at any hardware store.

Over the past three years, Corsi boxes have enjoyed major community uptake and an open source design evolution largely documented on Twitter (now X). These DIY designs have been accompanied by community CO_2 measurement education (see *Make:*'s "Plan CO_2" series, makezine.com/?s=plan+co2) and new open source databases about IAQ in public spaces.

Our organization, Open Source Medical Supplies (osms.li), has been diligently curating the Corsi box research and design ecosystem into our project library. Here is a simple breakdown of current Corsi box designs, their build cost, and user tools to evaluate the best project for your indoor setting.

WHY DO I NEED A CORSI BOX?

Clean indoor air is an important element of public health. Exposure to airborne pathogens and particles such as dust, smoke, mold, and pollen can negatively impact heart, lung, and brain function, and spread infectious disease. In rooms with poor ventilation, harmful particles can linger for hours or even days without an effective air purifier. The large cleaning surface area of MERV-13 filters makes Corsi boxes a more thorough and rapid purifying solution when compared to some commercial units (Figure Ⓐ).

HOW DOES IT WORK?

A Corsi box draws in contaminated air through the filter sides, and emits clean air through the

Festucarubra CC BY-SA 4.0, Matthew Azevedo, Amara Holder, Hannah Halliday, Larry Virtaranta, Amanda Hu CC BY-SA 4.0

CADR (CFM)

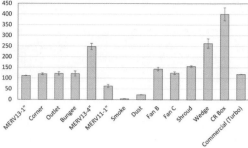

Ⓐ Clean air delivery rate (CADR) of various DIY air cleaner designs, measured by researchers at U.S. Environmental Protection Agency.

Ⓑ Basic Corsi-Rosenthal cube design: Dirty air is drawn through four filters around the sides, clean air exits the top.

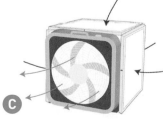

Ⓒ Alternate Corsi-Rosenthal cube design: Dirty air is drawn through four filters, clean air exits one side.

D Proposed Non-infectious Air Delivery Rates (NADR) for Reducing Exposure to Airborne Respiratory Diseases; The Lancet Covid-19 Commission Task Force on Safe School, Safe Work, and Safe Travel

	Volumetric flow rate per volume	Volumetric rate per person		Volumetric flow rate per floor area	
	ACHe	cfm/person	L/s/person	cfm/ft^2	L/s/m^2
Good	4	21	10	0.75 + ASHRAE minimum outdoor air ventilation	3.8 + ASHRAE minimum outdoor air ventilation
Better	6	30	14	1.0 + ASHRAE minimum outdoor air ventilation	5.1 + ASHRAE minimum outdoor air ventilation
Best	> 6	> 30	> 14	> 1.0 + ASHRAE minimum outdoor air ventilation	> 5.1 + ASHRAE minimum outdoor air ventilation

top (Figure **B**). This clean air then circulates through the room while the purifier continually cleans more air. An alternate design emits clean air through one side (Figure **C**).

HOW DO I MEASURE THE EFFECTIVENESS OF A DIY PURIFIER?

Clean air delivery rate (CADR) describes how much clean air the purifier is emitting. This measurement, combined with a given room size, determines *air changes per hour (ACH)*, which defines how many times per hour the volume of air in a given space is replaced with recirculated air. The Lancet Covid-19 Commission recommends a minimum of 4 to greater than 6 ACH to reduce exposure to airborne respiratory diseases (Figure **D**).

WHY CAN'T I JUST USE A COMMERCIAL AIR CLEANER?

You can. Clean Air Stars' open source calculator at filters.cleanairstars.com can help you pick an adequate commercial air purifier. DIY air purifiers are sometimes more affordable, more portable, and have a greater cleaning filter surface area when compared to some commercial units, especially during wildfire smoke events. They also make great STEM classroom projects!

BOX FAN CORSI BOXES
ORIGINAL CORSI-ROSENTHAL CUBE

Time Required: 30–60 Minutes
Difficulty: Easy
Cost: Under $100

The "OG" Corsi-Rosenthal Cube is built with a 20" box fan, duct tape, and four 20" MERV-13 filters (Figure **E**).

Shiven Taneja has provided excellent build guides in English (makezine.com/go/corsi-how-to), French, German, and Spanish, including a mini-Corsi box and UK-specific instructions (makezine.com/go/more-corsi-how-tos). These include a cardboard shroud that narrows the fan flow slightly to reduce backflow; you can make it for free, from the box the fan came in.

ⒺHOW TO BUILD A CORSI-ROSENTHAL BOX

The Corsi-Rosenthal Box is an affordable DIY air-cleaning system made with simple materials found in hardware stores. The box fan pulls air through the filters on the sides and blows out clean air. It is proven to reduce indoor exposure to airborne particles, including those containing the virus that causes COVID-19. The box can also decrease the levels of other particles in the air, such as dust or wildfire smoke.

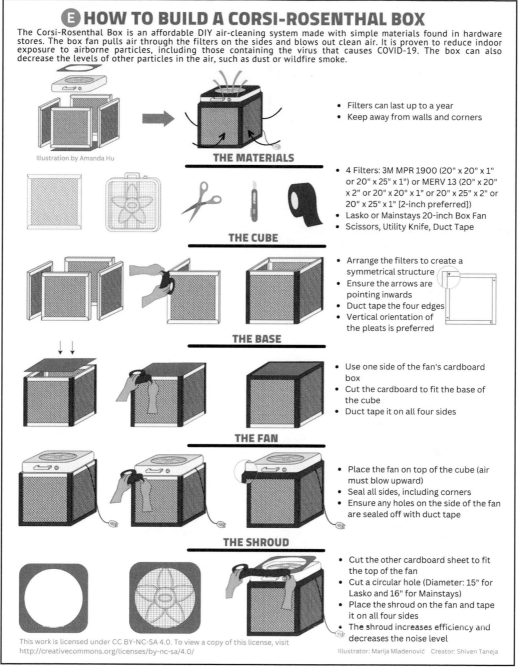

Illustration by Amanda Hu

THE MATERIALS

- Filters can last up to a year
- Keep away from walls and corners

THE CUBE

- 4 Filters: 3M MPR 1900 (20" x 20" x 1" or 20" x 25" x 1") or MERV 13 (20" x 20" x 2" or 20" x 20" x 1" or 20" x 25" x 2" or 20" x 25" x 1" [2-inch preferred])
- Lasko or Mainstays 20-inch Box Fan
- Scissors, Utility Knife, Duct Tape

THE BASE

- Arrange the filters to create a symmetrical structure
- Ensure the arrows are pointing inwards
- Duct tape the four edges
- Vertical orientation of the pleats is preferred

THE FAN

- Use one side of the fan's cardboard box
- Cut the cardboard to fit the base of the cube
- Duct tape it on all four sides

THE SHROUD

- Place the fan on top of the cube (air must blow upward)
- Seal all sides, including corners
- Ensure any holes on the side of the fan are sealed off with duct tape

- Cut the other cardboard sheet to fit the top of the fan
- Cut a circular hole (Diameter: 15" for Lasko and 16" for Mainstays)
- Place the shroud on the fan and tape it on all four sides
- The shroud increases efficiency and decreases the noise level

Illustrator: Marija Mladenović   Creator: Shiven Taneja

Pros & Cons: Box fan Corsi models are the cheapest and easiest-to-build DIY air purifiers, with an easily sourced BOM and short build time. They are the quickest to deploy in an airborne contaminant emergency, such as wildfire smoke.

However, when operated on the highest setting the box fan Corsi can be loud, causing it to be turned off in high-risk indoor areas. The louder the fan, the more likely it is to be turned off. The most effective cleaner is the one that's actually running.

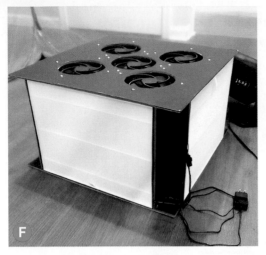

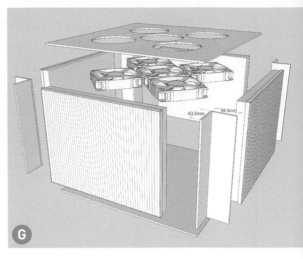

PC FAN CORSI BOXES

As Corsi boxes became more common, researchers identified a demand for quieter models with a lower decibel (dB) rating to encourage people to keep the purifier running in settings where transmission risk is highest, like indoor rooms with multiple people, including classrooms. This need, along with the preference for a more compact design, catalyzed the development of versions using PC fans.

WISSMAN CORSI CUBE
Time Required: 2–3 Hours
Difficulty: Moderate
Cost: $140–$150

One of the earliest and most popular PC fan Corsi boxes was created by Rob Wissman using five PC fans and four Ikea Starkvind filters (Figures **F** and **G**), and shared on GitHub (github.com/robwiss/diy_air_purifier/blob/main/5_Arctic_P14_4_Starkvind).

Wissman's cube-shaped build inspired the most popular current models — the compact, portable, two-filter rectangular PC fan air cleaners. Due to their small floor footprint and almost-inaudible fan noise, PC fan designs continually clean the air without creating auditory disruptions. Joey Fox, IAQ Advisory Chair for the Ontario Society of Professional Engineers (OSPE), has curated several of these designs, all featuring detailed build instructions with diverse frame materials.

AZEVEDO CORSI BOX
Time Required: 2–3 Hours
Difficulty: Moderate
Cost: $150–$220

Check out this slim two-filter Corsi box built by Matthew Azevedo, who has shared complete build instructions at It's Airborne (itsairborne.com/pc-fan-corsi-rosenthal-guide-a611dabf7e0c). A remix of the Wissman build, it features six P14 fans and two Ikea Starkvind filters for quiet operation and sleek design (Figures **H** and **I**). To save money, you can use five fans and downgrade them to the quieter P12 type.

CORSI KITS
CLEAN AIR KITS STEM PANEL KIT
Time Required: 30 Minutes
Difficulty: Easy
Cost: $174–$309

Clean Air Kits offers DIY STEM kits for home or classrooms (Figure **J**), including PC fans and all necessary components except for the filters, which the purchaser must source. A filter selection guide is provided on their site at cleanairkits.com/pages/stem-kits.

BREATHE FREE!

With your DIY Corsi box air filter you're ready for the next wildfire or Covid outbreak. Or if you wish, you can run it all the time to upgrade IAQ in your home or workplace.

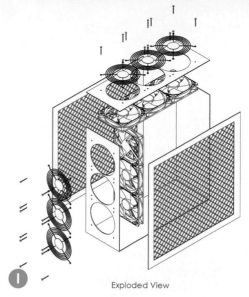

Exploded View

FILTERS AND WILDFIRE

Indoor air filters typically last months, but the EPA recommends more frequent filter changes for air purifiers exposed to wildfire smoke. The filters should be replaced when they are dark-colored or smell like smoke. ◐

HELPFUL LINKS

- **Clean Air Stars** — open source database for commercial air purifier selection and public space IAQ: cleanairstars.com/filters

- **OSMS Air Movement & Climate Control** — see additional air purifier designs: osms.li/Air-Movement-and-Climate-Control

- **CADR Calculator** — determine your Clean Air Delivery Rate (CADR) and Air Changes Per Hour (ACH) required for a given room size: reviewsofairpurifiers.com/cadr-calculator

- **It's Airborne** — layperson-accessible IAQ guides on Medium: itsairborne.com

- **Clean Air Advocates** — toolkit for clean air in schools presentations: lieslmcconchie.com/clean_air_advocates

- **IAQ for wildfire smoke** — tips from the EPA: epa.gov/air-research/research-diy-air-cleaners-reduce-wildfire-smoke-indoors

- **Proposed Non-infectious Air Delivery Rates (NADR) for Reducing Exposure to Airborne Respiratory Infectious Diseases** — The Lancet Covid-19 Commission Task Force on Safe Work, Safe School, and Safe Travel: tandfonline.com/doi/full/10.1080/02786826.2023.2249963

- **Efficacy of Do-It-Yourself air filtration units in reducing exposure to simulated respiratory aerosols** — Science Direct: sciencedirect.com/science/article/pii/S0360132322011507

CHRISTINA COLE (San Francisco) is a founding member and head of documentation for Open Source Medical Supplies (OSMS), and serves on the steering committee for the Global Open Source Quality Assurance System (GOSQAS). She is also a co-founder and director at the REAP Center in Alameda, California, and is pursuing a finance degree from Johnson and Wales University.

VICTORIA JAQUA (Amarillo, Texas) is a founding member of OSMS and co-author of "Design | Make | Protect," an OSMS and Nation of Makers (NOM) white paper documenting the PPE distributed manufacturing response to the Covid-19 pandemic in 2020. She is a curator for the OSMS Project Library and also serves on the GOSQAS steering subcommittee.

Rob Wissman, Matthew Azevedo, Abrar Shakib, Clean Air Kits

The Soft Matrix

Written and
photographed
by Lee Wilkins

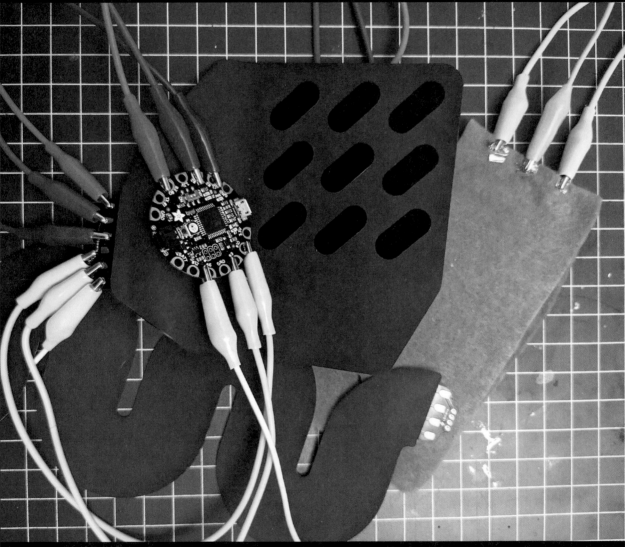

LEE WILKINS is an artist, cyborg, technologist, and educator based
in Montreal, Quebec, a board member of the Open Source Hardware
Association, and the author of this column on technology and the
body and how they intertwine. Follow them on Instagram @leeborg_

Make soft touch pads and panels in conductive fabric to control wearables, LEDs, and more

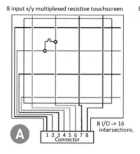
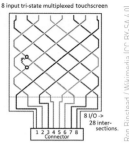

8 input x/y multiplexed resistive touchscreen

8 input tri-state multiplexed touchscreen

8 I/O -> 16 intersections.
1 2 3 4 5 6 7 8 Connector

8 I/O -> 28 inter-sections.
1 2 3 4 5 6 7 8 Connector

Ron Binstead / Wikimedia (CC BY-SA 4.0)

A

Let's talk about the matrix! In this article, we're going to learn how to make and read touch-sensitive panels using matrices and conductive textiles. This might sound a bit intimidating, but we'll take it step by step and understand the basic concepts. These touch panels can be used with a microcontroller to control all kinds of projects, or with your computer!

First of all, matrices aren't that complex. A matrix is just a series of rows and columns — basically a table. You can pinpoint a location in the matrix by finding the intersection of rows and columns, for example location 2b in Table 1.

TABLE 1: SIMPLE COORDINATE MATRIX

	1	2	3
a	1a	2a	3a
b	1b	2b	3b
c	1c	2c	3c

This concept is used widely in touch pads and panels, because you can easily use it to figure out where a user's fingertip is physically located using the coordinate system. These days, most touchscreens use *capacitive touch* or other modern tech to distinguish greater detail, sensing the fingertip by tiny changes in electrical field.

But some of the earliest touchscreens, like the Nintendo DS, use *resistive touch* technology to sense the fingertip by pressure: conductive rows and columns of a grid are separated by a semi resistive barrier, so that when they're pressed together, a row and column are connected and you can pinpoint that place on the grid (Figure **A**). In a manufactured resistive touchscreen, a transparent conductive material called ITO (indium tin oxide) is applied in a very thin and

dense grid to the screen panels so the screen is still visible.

You can also find matrices as common components in DIY electronics, such as an LED square matrix, 7-segment display, or button array. While a touchscreen works by keeping conductive traces apart to sense coordinates, an LED matrix works by sending coordinates from a computer program — connecting the positive and negative ends of an LED through rows and columns to light up a specific light. I really like this matrix concept, because it can be used to make really interesting and complex interactions with super simple tech!

There are touch matrices everywhere — think of every screen to buy groceries or transit tickets, or to use your phone, computer, or tablet. But resistive touch technology is a bit out of date, so you'll probably only encounter it in more frustrating scenarios. Modern touch technology uses capacitive touch, so it can have multi-touch inputs and much more precision.

In order to make a "soft" touch panel, we're going to use a few concepts that I explored in my previous columns. Like our soft pressure sensor from Volume 86, you'll be using both resistive and conductive materials to make a "sandwich." (I'm continuously realizing that most electronics projects I share here are described as sandwiches!)

Essentially, you'll arrange conductive traces in rows and columns which are separated by a semi-conductive material. When the intersection of the rows and columns is pressed, the resistance between the traces changes, and you can detect the physical location of the touch. Depending on the density of your grid, you'll get more or less accuracy. Let's start simple.

MAKE YOUR RESISTIVE TOUCH MATRIX

What I love about making your own touch matrices is that you can make them any shape you want. Try exploring wobbly lines, new textile materials, or expanding and contracting the grid size. Even though most resistive matrices have a uniform grid, there's no reason you can't get weird with it and make an undulating, intersecting pattern with varying degrees of density!

1. PREPARE YOUR MATRIX

You can either use a pre-made touch matrix, like Loomia's Mega Pressure Matrix (Figure **B**), or make your own. Loomia makes a variety of amazing soft surface interfaces for prototyping. Their touch matrix panel is made of a sandwich of materials similar to what we will make here, but prefabricated and very reliable. The back of the panel also has a super handy adhesive coating to mount onto anything. And it's got big connection pads for easy prototyping with alligator clips, with matching through-hole pads you can solder permanently when you're ready (Figure **C**).

There are a series of other cool Loomia tools, like connectors, buttons, and even a heating panel (loomia.com/shop), but that's for another article.

If you're making your own matrix, first you'll have to design your grid. An easy way to start is a regular grid with about 2" spacing between rows and columns. Like most DIY electronics, especially textiles, there's always the temptation to go small for the first try, but giving yourself some space will make the process easier.

Lay out the rows on one piece of material and the columns on another. Make sure they are straight and overlap properly if you place them on top of each other. It's helpful to make a template with a marker first, and lay your traces on top (Figure **D**). Let's start with a 3-row by 3-column piece, making 9 intersections.

Make sure the rows and columns go to the edge of your material on at least one side. Place a piece of velostat (Figure **E**) fully separating both grids (Figure **F**) and then sew it together with nonconductive thread into a delicious matrix sandwich. Rows and columns should never touch.

Attach an alligator clip to one end of each

TIME REQUIRED: 1–2 Hours

DIFFICULTY: Easy

COST: $40–$60

MATERIALS
FOR THE TOUCH MATRIX:
» **Not-stretchy fabric, Velostat conductive plastic sheet, and copper tape**
    —OR—
» **Loomia Mega Pressure Matrix or Mini Pressure Matrix**

» **Adafruit Flora microcontroller** Adafruit 659, or similar Arduino compatible
» **Alligator clip leads (6)**
» **Jumper wires (6)**
» **LED strip (optional)** WS2812 or similar

TOOLS
» **Scissors**
» **Needle and thread and/or sewing machine**

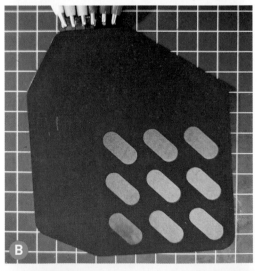

B

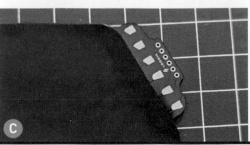

C

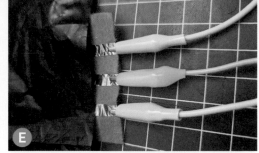

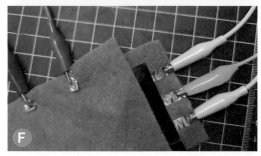

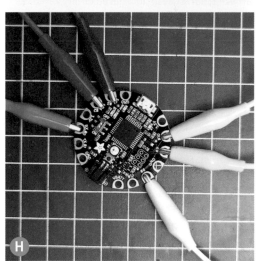

row and column. Connect the opposite clip to a jumper wire that you'll connect to your Arduino. If you're using a Loomia matrix, you can cover the through-hole pads with tape (or snip them off) and attach 6 alligator clips (Figure **G**).

To wire the circuit, you must connect each row and column correctly. We'll make all the rows Analog, and all the columns Digital. I've used an Adafruit Flora because it's made to attach alligator clips, and it's round which is great for soft circuits (Figure **H**). But this concept will work with any similar microcontroller; just check your pinout diagram to find digital and analog pins (Figure **I**).

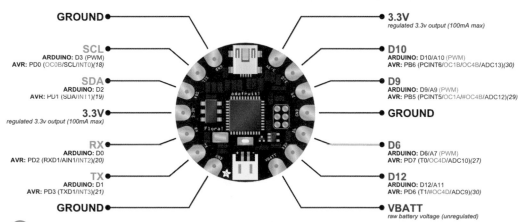

GROUND

SCL
ARDUINO: D3 (PWM)
AVR: PD0 (OC0B/SCL/INT0)(18)

SDA
ARDUINO: D2
AVR: PD1 (SDA/INT1)(19)

3.3V
regulated 3.3v output (100mA max)

RX
ARDUINO: D0
AVR: PD2 (RXD1/AIN1/INT2)(20)

TX
ARDUINO: D1
AVR: PD3 (TXD1/INT3)(21)

GROUND

3.3V
regulated 3.3v output (100mA max)

D10
ARDUINO: D10/A10 (PWM)
AVR: PB6 (PCINT6/OC1B/OC4B/ADC13)(30)

D9
ARDUINO: D9/A9 (PWM)
AVR: PB5 (PCINT5/OC1A/#OC4B/ADC12)(29)

GROUND

D6
ARDUINO: D6/A7 (PWM)
AVR: PD7 (T0/OC4D/ADC10)(27)

D12
ARDUINO: D12/A11
AVR: PD6 (T1/#OC4D/ADC9)(30)

VBATT
raw battery voltage (unregulated)

FLORA Wearable Electronics Platform adafruit.com/products/659
drawing 2012 by J. M. DeCristofaro -- CC-BY-SA 3.0

Adafruit

2. SET UP YOUR CODE

You can find the full, commented Arduino code at github.com/LeeCyborg/Resistive-Touch-Matrix. I've adapted and modified this code from Kobakant's original resistive touch concept at kobakant.at/DIY/?p=7443. It's a bit hard to parse, so let's break it down.

First, we will initialize variables we're going to use and define the number of rows and columns so Arduino knows how to cycle through them.

```
#define numRows 3
#define numCols 3
```

Next we add the names of the pins to our **rows[]** and **cols[]** arrays — make sure to consult the pinout diagram! — and then create a two-dimensional array to hold the incoming variables so we can read the whole matrix at once:

```
int rows[] = {A10, A9, A7};
int cols[] = {3, 2, 0};
int incomingValues[numRows][numCols] = {};
```

In our **setup** function, we cycle through all the rows and set them to **INPUT_PULLUP**.

> **NOTE:** If your controller doesn't have an internal pullup resistor, you can add an external one (roboticsbackend.com/arduino-input_pullup-pinmode).

We will make all of the columns **INPUT** as well and start the serial monitor:

```
void setup() {
    for (int i = 0; i < numRows; i++) {
        pinMode(rows[i], INPUT_PULLUP);
    }
    for (int i = 0; i < numCols; i++) {
        pinMode(cols[i], INPUT);
    }
    Serial.begin(9600);
}
```

This is where the code gets a bit tricky. We're going to fill every element in the second array by cycling through rows and then columns. We can see what column we're in because we've set it to **OUTPUT**, and then cycle through all the rows inside of it. This way we don't need as many pins, because we are isolating each row.

Upload this code (Figure **J**) to your Arduino and test it out! Have a look in the serial monitor to see the output.

3. TEST YOUR PRESSURE TOUCH MATRIX

At the end of the code are two functions that can help us debug. You'll notice a grid of numbers in the serial monitor. The function **printGrid** cycles through all the rows and columns and prints them here as a number. When you push a button, or a copper tape intersection (Figure **K**), the value of that entire row will be lower, but the exact button location will also be clear as the lowest value.

```
void loop() {
    // On every loop, iterate through the columns
    for (int col = 0; col < numCols; col++) {
        // Set the entire column to output OUTPUT
        // and write a LOW signal to it.  This lets us know what column we
        // are reading.
        pinMode(cols[col], OUTPUT);
        digitalWrite(cols[col], LOW);
        for (int row = 0; row < numRows; row++) {
            /// In that column, read the row.
            incomingValues[row][col] = analogRead(rows[row]);
        } // end row
        // Then set it back to input so we can look at the next column
        pinMode(cols[col], INPUT);
    } // end col

    printGrid();
    findActiveButton();
    delay(10);
}
```

In order to find the lowest value, we can use the **findActiveButton();** function (Figure ) . This function cycles through all the incoming values and determines which is the lowest. If any of them are under **100**, then it's likely a button is being pressed. If you made your own touch matrix, it's possible these numbers are different. You can use a calibration function to determine a baseline value (see Volume 86, pages 96–97), or just check manually what the values tend to be when something is pressed.

If you're getting erratic readings, it's possible that your resistive material isn't fully separating your conductive rows and columns, or that a connection isn't being made somewhere. Double-check your sewing and wiring.

To give your touch matrix a different feel, you can experiment with placement and even integrate the soft pressure sensors I shared in Volume 86, "Squish It! How to make a DIY pressure sensor."

INTEGRATE AND INSTALL!

Now that you're able to read your resistive matrix, you can make it do things! There are a number of ways of doing this.

- You can use the different locations on your touch matrix to change an LED strip's patterns or colors, or to trigger a different interaction.
- You can connect your touch matrix to a computer and control a web page, game, or interactive art.
- You can use platforms like Processing to read the values through the serial monitor and trigger actions there.

It's fun to integrate these soft matrices into clothing that you can control in an organic way. I've mounted my Loomia matrix to my bag so I can control my wearable LEDs! ◖

```
void findActiveButton() {
    // Find the smallest value.
    int smallest = 1024; // Set to max so we know if it isn't working
    int smallestRow = -1;
    int smallestCol = -1;
    for (int col = 0; col < numCols; col++) {
        for (int row = 0; row < numRows; row++) {
            // Iterate through rows and columns, check each value if it is
            // smaller than the smallest, then make it the new smallest if it is.
            // and set the rows and columns to the current position
            if (incomingValues[row][col] < smallest) {
                smallest = incomingValues[row][col];
                smallestCol = col;
                smallestRow = row;
            }
        }
    }
    if (smallest < 100){
        // If there is a button being pressed it will be less than
        // 100, so you can identify it.
        Serial.print("There is a button being pressed in row: ");
        Serial.println(smallestRow);
        Serial.print("and column: ");
        Serial.println(smallestCol);
    } else {
        Serial.println("Nothing is being pressed");
    }
}
```

Change Sensor

Find coins and other treasure with a DIY metal detector circuit
Written by Fredrik Jansson with Charles Platt

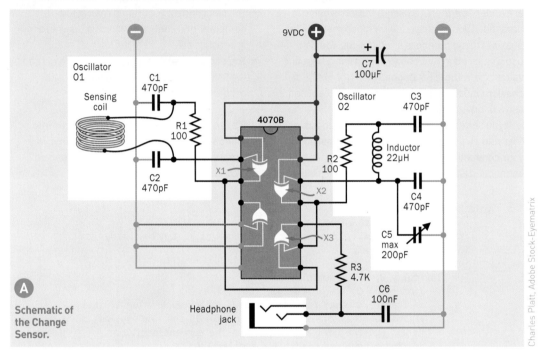

9VDC

Oscillator
O1

Sensing
coil

C1
470pF

R1
100

C2
470pF

X1

4070B

C7
100µF

Oscillator
O2

C3
470pF

R2
100

Inductor
22µH

X2

C4
470pF

X3

C5
max
200pF

R3
4.7K

C6
100nF

Headphone
jack

Schematic of
the Change
Sensor.

Charles Platt, Adobe Stock-Eyematrix

People don't carry change in their pockets anymore. Or do they?

With the Change Sensor, you can find out. This ultra-simple circuit will also search for hidden treasure — or your house keys, if you drop them in some tall grass. In fact, any metallic object that conducts electricity may reveal itself to you, for this is a basic metal detector circuit.

The schematic is shown in Figure A, and the breadboard layout is in Figure B. The symbols labeled X1, X2, and X3 inside the 4070B chip are XOR logic gates, three of which are active.

COMPONENTS
The parts on our shopping list are low in price and easily obtained. If you search online for a "200pF

CHARLES PLATT is the author of the bestselling *Make: Electronics*, its sequel *Make: More Electronics*, the *Encyclopedia of Electronic Components Volumes 1–3*, *Make: Tools*, and *Make: Easy Electronics*. makershed.com/platt

FREDRIK JANSSON is a researcher in physics and weather modeling, and coauthor of the *Encyclopedia of Electronic Components Volumes 2 and 3*.

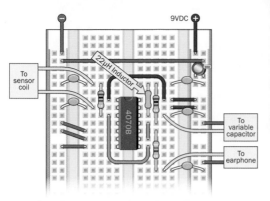

B Breadboard layout. See schematic for component values.

TIME REQUIRED: **1–2 Hours**
DIFFICULTY: **Easy**
COST: **$10**

MATERIALS
» **Breadboard and jumper wires**
» **4070B quad XOR chip**
» **Resistors: 100Ω (2) and 4.7kΩ (1)**
» **Inductor, 22µH** such as Bourns 78F220J-RC
» **Ceramic capacitors: 470pF (4) and 100nF (1)**
» **Electrolytic capacitor, 100µF**
» **Variable capacitor with screw terminals and tuning wheel, 200pF** such as Model 223P; avoid 223F, which is much smaller. (May be described as 140pF + 60pF. See text.)
» **Generic earphone(s)** with 3.5mm mono or stereo jack plug
» **Earphone connection block** See text for details.
» **Hookup wire, solid, 22 gauge (13 feet)**
» **Terminal block, "Euro style," with 3 pairs of screw terminals** ideally spaced ⁵⁄₁₆" (8mm)
» **9V battery and battery connector**
» **On-off switch (optional)**

TOOLS
» **Wire strippers**
» **Pliers**
» **Small flat-blade screwdriver**
» **Adhesive tape**
» **2-liter bottle** for forming your coil

This project is adapted from our forthcoming book *Make: Radio*, available for pre-order now at the Maker Shed, makershed.com.

C The variable capacitor recommended for this project.

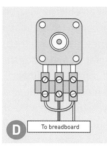

D Connecting the variable capacitor through a terminal block.

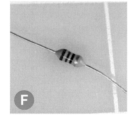

E A plastic wheel should be added to your variable capacitor.

F An inductor is easily mistaken for a resistor.

variable capacitor," one of the first hits is likely to be the component we suggest, part number 223P (Figure C). It actually contains two variable capacitors on the same shaft, with maximum values of 140pF and 60pF (picofarads). You need to connect them in parallel, for a total of 200pF. You can use alligator jumper wires for this, but a terminal block with screws spaced at ⁵⁄₁₆" (8mm) is better (Figure D). The center terminal should go to negative ground on your breadboard.

Wherever you buy your variable capacitor, also buy a plastic tuning wheel (Figure E) which screws onto the shaft, to prevent the touch of fingers from affecting the capacitance.

The inductor in our list (Figure F) can be found at any large electronics supplier. It looks like a resistor, but inside it is a coil of very fine wire. Its value is expressed in microhenries, written as µH.

The circuit will signal you via an earphone. If you can find a high-impedance type, often sold "for crystal sets," that's ideal; but any earphone(s) terminating in a 3.5mm jack plug will work. See Figure G on the following page.

To connect the plug to your breadboard, search online for a "3.5mm jack terminal block connector." Its screws may be identified as T, S, and R (for the tip, sleeve, and ring of a stereo plug), or L, R (left and right stereo channels), and a ground symbol.

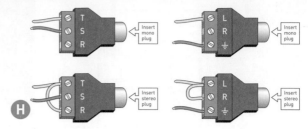

You can use any earphone with a 3.5mm mono plug (left) or stereo plug (right).

How to wire different types of audio adapter.

Figure **H** shows how to connect either type of adapter when using stereo or mono headphones. The blue wire goes to negative ground in each case, while the yellow wire carries the signal.

The yellow sensor coil shown in Figure A can be adjusted to suit your interests. A smaller-diameter coil will be more sensitive to small objects such as coins. A larger coil will respond to large objects. In Figure **I**, the large coil is 4¼" diameter, created by wrapping 10 turns of 22-gauge hookup wire around a 2-liter bottle of soda. The smaller coil contains 17 turns, 2" in diameter. Try them one at a time.

SEEKING A SIGNAL

After you connect your coil and attach a 9V battery, turn the variable capacitor until you hear a sudden whistling sound. Now turn the capacitor very slowly until the sound goes lower in pitch and finally disappears. Keep turning, and it should reappear and rise in pitch. Back up and set the capacitor at the silent spot between the falling and rising frequencies.

Now move a metal object over or into the coil, and if the object is similar in size with the coil, the circuit will whistle in response.

What if you hear nothing at all? Add a turn of wire to your coil, or subtract a turn.

THE CONCEPTS

If you're wondering how this simple circuit magically detects metal, the answer involves capacitance and inductance.

Capacitance measures the ability of an object to store extra electrons. The capacitors in a circuit can do this, and your body can do it, too. If you ever feel a little shock when

touching a metal object such as a door handle, your body just discharged some electrons which it had accumulated relative to the environment.

In an electronic circuit, when you apply voltage to a capacitor that has no charge, it sucks up some current as electrons accumulate on one of its plates. Gradually the current diminishes until the capacitor is fully charged.

Inductance is the ability of electric current in a wire to induce a magnetic field. This effect is multiplied when the wire is coiled. When you apply voltage to a coil, initially the current seems to encounter resistance, because energy is being taken to create the magnetic effect. After the field is stable, when you disconnect your power supply, there is a sudden reversed surge of current that is created by the field collapsing.

The two components seem to have opposite personalities. The capacitor sinks a surge of current and then gradually blocks it; the inductor blocks a surge of current and then gradually lets it flow. What would happen if you put a capacitor and an inductor in parallel, as in Figure **J**?

If you use the switch to supply a brief pulse of current, initially the coil blocks the flow while the capacitor acquires a charge. Then if you open the switch, the capacitor discharges through the coil. When the flow ceases, the magnetic field collapses, sending current back to the capacitor and recharging it with opposite polarity. At the end of that cycle, the capacitor discharges again, and the sequence repeats. This is sometimes known as a *tank circuit* because current is sloshing around like water in a tank. More accurately, it is an *oscillator*.

Figure **K** shows the voltage in

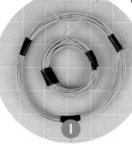

The 2" coil contains 17 turns of wire. The 4¼" coil contains 10 turns.

Charles Platt

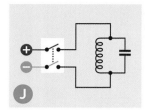

The concept of an oscillator using a coil and a capacitor in parallel.

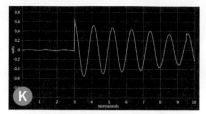

Decaying oscillations in a tank circuit after a brief pulse is applied.

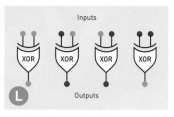

Inputs and outputs using an XOR logic gate.

the circuit in Figure J after the power is pulsed very briefly. The circuit has a small amount of resistance, so the current gradually loses energy in the form of heat, and the oscillations die away.

But in our circuit, logic gates X1 and X2 feed back just enough power into O1 and O2, through resistors R1 and R2, to keep the oscillations going. The frequency is then determined by capacitors C1 through C5, the value of the inductor, and the diameter and number of turns of the sensing coil.

Oscillators are fundamental in radio, when you want to receive a station or change the frequency of a transmission. We have been writing a book titled *Make: Radio* which will be out in 2024, in which we go into the topic in much more detail.

MIXING OSCILLATORS

When you apply voltage to a coil, it can induce tiny amounts of current in any nearby object that conducts electricity, and this changes the frequency of the circuit containing the coil. Because metal objects are conductive, the coil will react if it senses some metal nearby.

The only problem is that if we choose a variable capacitor and inductor that are conveniently and affordably small in size, the circuit will oscillate at a frequency that is much too high to hear. We can get around this limitation by enabling you to hear the difference between O1 and O2.

Remember how an XOR gate works. Its output is low when both inputs are the same, and high when the inputs are not the same, as shown in Figure **L**, where blue dots indicate 0 volts and red dots indicate supply voltage.

In our circuit, you begin by setting O1 and O2 to the same frequency, around 2MHz; this occurs when you hear silence in your headphones. Now any disturbance will cause O1 and O2 to go out of phase, thousands of times per second — at an audio frequency.

X3 mixes the signals from O1 and O2 before passing its output through R3 (which protects the logic chip from delivering excessive current). Capacitor C6, with R3, functions as a **low-pass filter:** it grounds high frequencies and only allows audio frequencies to reach the headphone jack.

When O1 and O2 are almost synchronized, the output from X3 is "off" most of the time. But when O1 and O2 are out of sync, the output from X3 is mostly "on," as shown in Figure **M**. The small fluctuations are eliminated by the low-pass filter, but the transition between "mostly off" and "mostly on" is audible in the earphone. You can add an audio amplifier such as an LM386 if you want the sound to emerge from a small speaker.

Maybe you've seen security personnel at airports waving a wand around, searching for metal objects under people's clothes. Our Change Sensor is not as sensitive as that, because it only uses a handful of components (Figure **N**). But the principle is much the same. ◎

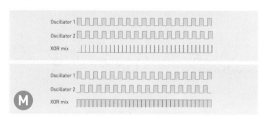

How an XOR gate combines oscillations that go in and out of sync.

The assembled circuit.

Very Sucky

Print a portable mini fume extractor, and solder more safely anywhere

Written and photographed by Chris Borge

A

Circuit Diagram

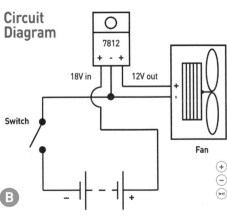

7812

+ - +

18V in   12V out

Switch

Fan

−   +

B

Marc de Vinck

Charcoal fume extractors of all shapes and sizes have been a staple project in makerspaces for many years. *Make:* even has its own mini fume extractor by Marc de Vinck (Volume 19) that's still up on the KipKay YouTube channel, now there's a throwback! Built into an Altoids tin and featuring a 40mm fan and a basic circuit, this little device was rough but functional (Figures Ⓐ and Ⓑ).

Since then, 3D printing has come into popularity and with it a much greater range of distinct designs appeared. I don't think I'd be wrong to assume that the rise of 3D printing has also bought with it a rise in casual solderers — makers who, like myself, don't stray much further in their electrical endeavors than adding a few LEDs to a project, and maybe a voltage step-down if I'm feeling bold.

Being a seldom solderer, a full-scale fume extractor just isn't something I have the space or money for. Thankfully modern non-leaded solder fumes are not as horrible as older lead-based solders, but breathing fumes is still not advisable. Initially I went with one of the cheap boxy fume extractors you get on eBay. While decently functional, this model was bulky (Figure Ⓒ), making it a pain to work around, and even worse if you have to film what you're doing (not an everyone problem of course, but it was annoying for me).

I wanted something small, convenient to set up, and stored in such a way that using it would almost be easier than not. Naturally, I turned to 3D printing. Freed from the need to force parts into less-than-ideal enclosures, 3D printed models can focus more on new features that would have been impractical or impossible without the new technology. Designs with built-in USB charging, articulated arms, and novelty enclosures are just some of what's available for free on sites like Thingiverse and Printables.

Despite the variety, I couldn't find the *perfect* one for me. The existing models used mostly fans around 100mm, but I wanted something even smaller: 40mm like the old Altoids tin designs used. This wasn't for nostalgia, but practicality. The helping hands I used had a third arm which I thought would be perfect to mount a small fume extractor on, but nothing really fit the bill. There were larger extractors on arms, and I could scale

TIME REQUIRED: A Weekend
DIFFICULTY: Easy
COST: $15–$20

MATERIALS
» **Mini computer case fan, 40mm** I used a 2-wire fan and soldered on a barrel jack to match a 12V power brick I had. If you use a USB fan, this project is no-solder.
» **Activated carbon filter, 10mm thick**
» **3D printer filament** in pink, black, and red

TOOLS
» **3D printer**
» **Soldering iron and solder (optional)**
» **Wire cutters/strippers (optional)**

CHRIS BORGE is a hobbyist designer and 3D printer in South Australia. He works in media and is studying for a marketing degree.

one down, but I would much rather release a custom solution.

Getting to brainstorming, I had a clear idea of what I wanted to make (Figure Ⓓ). It's really just a fan on an arm. My job was mostly to make it look pretty. During this brainstorming process I considered a range of design concepts and styles, many of which were influenced by the *Portal* video

game series. Looking into my sketchbook you can just about make out some of the design features of a Companion Cube, which was my number one option for a time. But then I had an idea: *What about Kirby?!*

A popular Nintendo character, Kirby is a pink blob who defeats enemies by inhaling them. The idea met essentially none of my original design goals but was just too perfect not to consider. A

quick Google showed me however that I wasn't the first; there were a few designs already available. I would've scrapped the idea here but looking at the files on Thingiverse, I saw room for improvement.

Not to say that any of these designs were bad, but they were all single-part prints, more like half a Kirby, since they print flat on their backs and need support for the space where the fan is inserted. They also needed a decent bit of painting to add extra features; again fine, but I prefer to print parts in different filament colors as it makes a good result more accessible.

I hopped onto Fusion 360 and got to designing. I stayed with the 40mm fan and 10mm charcoal filter as I had originally planned. Moving forward I was reminded once again that I really should learn Blender if I want to do more projects like this. Even creating basic "organic" shapes like Kirby's feet took a few tries, but I got a design I'm pretty happy with in the end (Figure E). The functional elements came together quickly but I spent a lot of time trying to match the proportions to images of the character to get it as close as possible.

As mentioned, the design prints in several parts. The arms and eyes both glue onto the main body and have mating holes to make correct installation easy (Figure F). The eyes have recessed portions to add pupils, with a dab of paint or even permanent marker. The feet glue onto a connecting piece which then slots into a dovetail in the back. This dovetail allows Kirby to tilt up and down to a more ideal angle (Figure G). Finally the fan is press-fit in place and the filter installed, and the two halves thread together without any hardware (Figure H).

I'd be lying if I said this wasn't one of my favorite things I've made. I almost released just Kirby, but I still wanted the functionality of the original idea. So I went through a similar process with the same hardware but in a very different, and admittedly much more functional, arm-mounted package (Figures I, J, and K). Interestingly I got a lot of comments saying it reminded them of GLaDOS, so it seems some of my initial *Portal* brainstorming made it through subconsciously.

These were really fun projects to work on and the community reaction was great. Again, these are the "better than nothing" option — if you solder a lot, proper extraction is recommended. But if you just want something small, the files are available for free on Printables (printables.com/@ChrisBorge/models) and you can learn more in my YouTube video (youtu.be/XuhWTwN0GvE). ◆

MORE SUCKY PROJECTS TO MAKE:

ALTOIDS TIN MINI FUME EXTRACTOR
The battery-powered mint tin version by Marc de Vinck from *Make:* Volume 19.
- makezine.com/projects/mini-fume-extractor
- youtu.be/klTfk6L7o6o

DIY FUME HOOD
Full-on fume management for painting and gluing, by Téa Forest from *Make:* Volume 80.
- makezine.com/projects/diy-fume-hood

CYCLONE DUST SEPARATOR
Capture the mess before it's made, with this cyclonic see-through dust collector by Raymond Mowder.
- makezine.com/projects/cyclone-dust-collector

Hack Your Toothbrush

Written by Cy Tymony

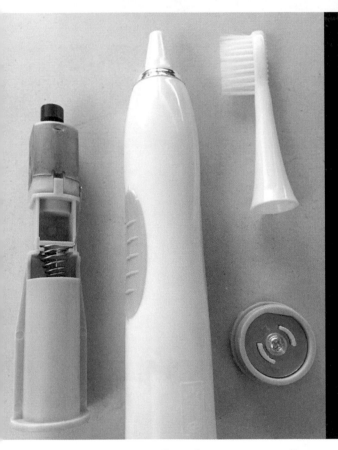

TIME REQUIRED: 20–30 Minutes

DIFFICULTY: Easy

COST: Free

MATERIALS
- » Discarded electric toothbrush, battery powered
- » Batteries, AA or 9-volt
- » Battery holder or clip
- » Old toy propeller (optional) or cut one out of a potato chip can lid
- » Soda bottles and potato chip cans
- » Cardboard
- » Adhesive foam tape
- » Electrical tape and transparent tape

TOOLS
- » Pliers
- » Small screwdriver
- » Scissors

Part 1: Salvage tiny motors from discarded electric toothbrushes and use them for fun projects!

CY TYMONY is a Los Angeles–based author and inventor who teaches super sneaky and simple ways to use everyday objects to make things. He's authored 10 volumes of his *Sneaky Uses* series, starting with the original, *Sneaky Uses for Everyday Things*, in 2003.

Several hundred million electric toothbrushes are manufactured, and sadly discarded, each year yet most people don't realize that toothbrush parts are salvageable and they don't have to wind up in a landfill or the ocean.

Using the magnets and DC motors inside discarded electric toothbrushes — and also inside motorized personal fans, pencil sharpeners, fabric shavers, facial brushes, and toy cars — I'll show you how to repurpose them to make 10 practical projects. Here in Part 1 we'll build five using motors, and in Part 2 in the next issue we'll build five using magnets!

TOOTHBRUSH DISASSEMBLY

These instructions are for the inexpensive electric toothbrush models that use replaceable AA batteries. Similar methods can be used for other varieties.

1. Remove the toothbrush end cap and battery.
2. Twist the case apart.
3. Pull the motor out.

MOTOR PARTS

The motors in inexpensive electric toothbrushes (and many toys and consumer devices) are permanent magnet Micro 150 or 130 models. They typically include a small weight on the end of the motor shaft so that it wobbles while it spins to vibrate the toothbrush. You can pry the weight off the shaft with a pair of pliers (Figure **A**).

To open the motor case, pry the 2 side tabs upward with a small screwdriver (Figure **B**).

Next, remove the end cap, which holds the motor contacts. Then wiggle and slide the rotor out of the case (Figure **C**).

The two curved magnets in the case are held in place by a V-shaped metal clip. Pry it out with pliers and a small flat blade screwdriver (Figure **D**).

HOW IT WORKS

Notice the motor parts labeled in Figure **E**. The two magnets inside the case (the *stator*) surround the *rotor* (also called the *armature*), which is a set of spinning electromagnets.

The rotor's armature wire coils receive power from the *commutator* — a ring with gaps in it — on the end of the rotor (Figure **F** on the following page). The commutator ring receives electrical

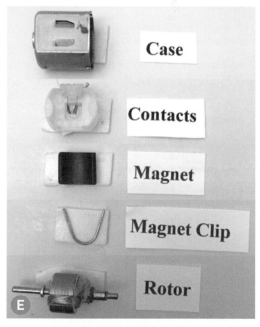

Case

Contacts

Magnet

Magnet Clip

Rotor

Cy Tymony and Bill Melzer

power from the end cap's contacts, which connect to the battery.

The commutator switches the polarity of the electromagnet armature coils so they alternately attract and repel the permanent magnets to keep the rotor spinning. To learn more, you can watch Jared Owen's animated electric motor tutorial at youtu.be/CWulQ1ZSE3c.

EASY MOTOR REPAIRS

Small motors usually fail because the wire coils are faulty, or the rotor does not spin freely, or the electrical contacts on the end cap do not properly contact the commutator ring.

If you have an inoperative motor:

1. Test the motor with a known-good power source.
2. Ensure that the rotor can spin freely.
3. Bend the contacts (Figure **G**) so they touch the commutator ring when in place (Figure **H**) and test it again.

MOTOR PROJECTS

Now you'll discover how to make five projects using a toothbrush motor. Ensure that the motor works with a fresh power source and has wire leads securely connected to its two contacts.

> **NOTE:** Check the motor's suggested voltage rating for use with your projects. Generally, the common Micro 150 motors can run fine with 3 to 6 volts. If you want to use a higher voltage power source the motor will run faster but the batteries will also drain faster. You can place a resistor in series with the battery to reduce voltage for longer power usage.

1. PERSONAL MINI FAN

By attaching a propeller from an old toy, or cutting one out of a chip can lid or soda bottle plastic, you can easily construct a toy fan with a battery clip and battery and adhesive foam tape as shown in Figures **I** and **J**. Connect the motor wires to the battery clip wires, and insulate the connections with electrical tape.

2. BUBBLE BLOWER

Position a wire loop dipped in bubble maker fluid in front of the propeller to turn your personal fan into a motorized bubble blower!

3. SNEAKY SPEAKER

A motor has wire coils and magnets inside, and so does an audio speaker. You can improvise and use a motor as a makeshift speaker in a pinch!

Cut one end of a ⅛" audio cable (Figure K) and attach its wires to the motor's wires.

Tape the motor securely to the bottom of a chip can. Then connect the audio cable plug to a sound source, like a radio (Figure L). Turn the radio volume up high and you will be able to hear the radio audio from the can.

How it works: The sound signal is causing the motor wires to become an electromagnet which is attracted to the permanent magnets. This vibrates the motor case and produces sound just like a speaker does!

4. MINI VACUUM CLEANER

Here's a novel project that will amaze your friends — convert your toothbrush into a mini vacuum cleaner!

Obtain a soda or water bottle that narrows in the middle. Cut off the top and bottom sections of the bottle, removing about 3" of the center section (Figure M), so that the top fits tightly over the bottom.

Poke holes around the base of the bottle. Then stick a piece of adhesive foam tape at the bottom inside.

Stick the bottle cap and motor to a chip can's plastic lid (Figure N). Place this motor assembly in the base of the bottle and lead the wires out the holes. Mount the battery outside the bottle with foam tape. Connect the motor and battery wires and insulate with electrical tape.

Place some screen material on the top of the base (Figure O) to prevent debris from getting into the bottom of the bottle. Press the top section of the bottle snugly over the base (Figure P).

Turn on the battery power supply and your toothbrush motor can now vacuum dust, small particles, pieces of paper, and more.

5. BONUS SCIENCE PROJECT

Demonstrate how a motor works while it's mounted outside its case! Remove the end cap and rotor from a motor. Mount the case and end cap backwards with the rotor between them on a piece of cardboard. Ensure the rotor can spin freely.

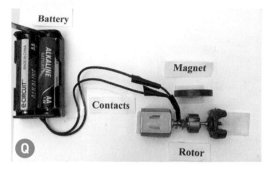

Battery · Magnet · Contacts · Rotor

Mount a strong magnet or two close to the rotor (or optionally, tape a strong motor underneath the rotor on the back of the cardboard). Press the battery wires on the commutator ring and the motor will spin (Figure Q).

GOING FURTHER

Small motors can provide many more possibilities for science and craft projects. Here are a few suggestions:

- Toothbrush motors can be used in innumerable science projects like a vibrobot toy, mini drill or sander, water pump, or electrical generator.
- Substitute salvaged toothbrush motors into inoperative devices and toys to prevent unnecessary discards.
- Copper wire in rotor armatures can be unwound and used for electrical circuits. Note that armature wire, aka magnet wire, has an enamel coating which must be gently sanded off at the ends to make electrical contact.
- Motor cases can hold small electronic parts, screws, etc.

UP NEXT: MAGNETS

If your motor is broken, you can still salvage the magnets inside to make practical projects. I'll show you how in the next issue of *Make:*! ⦿

Written and
photographed
by Becky Stern

Harvest Disposable
Vape Batteries

Give trendy products destined for e-waste a second life

How do vape companies get away with putting perfectly good rechargeable lithium batteries in a single-use device? And why does it seem like there are so many more of these around all of a sudden? It's ridiculous.

I think the shift has something to do with the Juul ban. Juul had a significant share of the market with its rechargeable, cartridge refillable device. When it was pulled from the U.S. market in 2022 by the FDA, new competitors popped up to fill the demand. In a market where brand recognition could mean a target on your back from regulators, companies aren't incentivized to make a refillable system. And selling you a completely new device every time the old one wears out is more profitable than selling you just a cartridge. Unless lithium becomes much scarcer, or it becomes a legal requirement to make them infinitely refillable, I don't see this trend changing anytime soon, unfortunately.

Some brands do include a charging circuit to make use of a larger liquid reservoir. But once the

BECKY STERN has authored hundreds of DIY tutorials about everything from microcontrollers to knitting. She is an independent content creator and STEM influencer living in New York City. Previously she worked as product manager at Instructables (Autodesk), director of wearable electronics at Adafruit, and senior video producer for *Make:*. She lives in Brooklyn and enjoys riding on two wheels, making YouTube videos, and collecting new hobbies to share with you.

liquid runs out, the device is still intended to be thrown out.

How did I get this many devices to take apart? Besides picking them up off the sidewalk, I asked my local Buy Nothing group and found a few willing folks who knew better than to throw these in the trash and were happy to offload their hoard.

LITHIUM BATTERY OVERVIEW

Lithium batteries are used in everyday devices like laptops, cell phones, hybrids, and electric cars. They are much lighter than traditional alkaline batteries and can last much longer. They can also be recharged multiple times — it's no wonder that they're in everything nowadays.

Lithium batteries contain layers of materials folded up together into a small shape. Energy is stored on either side of a battery "stack" and it wants to get from one side to the other, and circuits get power by making the charges do work along the way. As the battery discharges and produces an electric current, lithium ions are released by the anode to the cathode, causing a flow of electrons from one end to the other.

When charging, the opposite happens: the cathode releases lithium ions, which are then obtained by the anode. You can think of the act of charging like pushing the energy back to the other side so that it can be ready to go again. And if the layers are breached by, say, a puncture or by crushing, the battery can become shorted out, and the electrons get way too excited.

The *C-rate* is the measure of how quickly the battery can be discharged and recharged without damaging it. Failure modes include overcharging, over-discharging, and short-circuiting. Always use a charging rate appropriate for the battery's C rating, or a conservative guess.

TAKING APART THE DEVICES

Taking apart anything with a lithium battery in it is dangerous. You have to be careful not to damage the battery or short it out, or you could quickly have a concentrated fire hazard on your hands. So, don't do this at home without the supervision of someone who knows what they are doing!

I cracked open the cases using an awl and hammer to apply force to the seams in the plastic enclosures (Figure A).

TIME REQUIRED:
1 Hours
DIFFICULTY:
Moderate (Safety Concerns)
COST:
A Few Dollars
MATERIALS
» **Used vape dispensers** check your local Buy Nothing group
» **Adjustable LiPo battery charger** like SparkFun #PRT-14380 or DFRobot #DFR0668
» **JST wires**
» **USB power meter** like Pimoroni #TST001
» **Heat-shrink tubing**
» **Kapton tape** or electrical tape in a pinch
TOOLS
» **Awl** for cracking open the case
» **Hammer**
» **Gloves**
» **Pliers**
» **Wire cutters**
» **Soldering iron and solder**

Lithium Battery Do's and Don'ts

Here's a list of do's and don'ts when using LiPo batteries:

DO
• Charge your battery slowly and evenly, at a rate appropriate for its capacity
• Store your battery in a cool place
• Get a fireproof battery bag to keep your batteries in

DON'T
• Don't charge or discharge your battery too quickly or let it get too hot
• Don't leave your battery unattended while charging
• Don't discharge your battery below its minimum voltage level

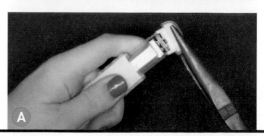

A

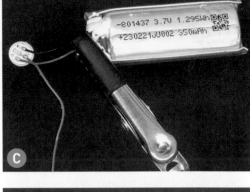

With the case open, the whole circuit and tank assembly usually slides right out. If it doesn't, you can reach in with pliers and pull on the plastic tank — not the battery or its wires.

I recommend wearing gloves when handling and disposing of the liquid tank, and give the battery a wipe before removing them. When I didn't, I had a hard time getting the vape juice stink out of my fingers even after washing my hands. This whole process is pretty smelly, and contact with skin or eyes can cause irritation and burning according to the FDA, so be careful.

Here are some examples of the batteries I found inside. These 280mAh (Figure **B**) and 350mAh batteries (Figure **C**) are about $5 retail, and perfect for wearables or other portable projects.

I desoldered them from the other circuitry, and then soldered their two wire connections to a new JST plug, with plenty of heat-shrink tubing to insulate the connections and relieve any strain on the tiny battery wires.

CHARGING THE BATTERIES

To charge the batteries up, you can't just connect them to power. You'll need a charger circuit with appropriate settings for the size of the battery. The charger monitors the battery and fills it up gradually and safely.

Some chargers have a DIP switch so you can toggle the charging rate, like Figure **D**. You should choose the fastest rate without going over the capacity of the battery. So, for example, for a 350mAh battery, I'd charge at 300mA. Others have solder pads you can bridge to set the charging rate, like Figure **E** — I chose 200 for the 280mAh batteries.

The large-capacity vape devices (Figure **F**) are nice for reusing because they come with a charging circuit. It's been interesting to see that charging chips vary quite a bit. Some have integrated overvoltage and overcurrent protection in the chip, but others just put a couple of PNP transistors on there and call it a day. Depending on the quality of the included charge circuit, you may wish to use it ... or lose it.

PROTECTION CIRCUITRY & SAFETY

The difference between reclaimed vape batteries and the nicer ones you might buy is the protection circuitry, or lack thereof. Lithium batteries can be damaged if they are drained or charged too much or too fast. Good hobby batteries will have a little circuit on them that cuts off the power when their voltage dips too low, and protects against shorts and dangerously high current with an overcurrent cutout. These vape batteries usually don't have any of that, so you need to add that circuitry yourself or leave the battery connected to a charger that has protection circuitry built-in, and is set to the proper charging rate for your battery.

To protect against accidental short circuits after the battery is out of the enclosure, I added some tape and heat-shrink to the exposed contacts (Figure **G**). Obviously, it's not a great idea to be adding any amount of heat to these batteries, so I was very judicious.

Electrical tape isn't the best kind, either. You should use high-temperature Kapton tape, but electrical tape is still better than just chucking all these bare batteries in a box together (Figure **H**).

When the device designers put these together, they knew the current draw would never exceed that of the air sensor and heater. But when you're reusing these batteries, thermal runaway could be a real concern — if your circuit draws more current than it is rated for (the conservative rule of thumb is the battery capacity per hour), you could start an unstoppable chain reaction inside the battery that causes it to explode. Remember the Samsung Galaxy Note 7?

In conclusion, I encourage you to pick these devices up if you see them littered, and at least get them to proper e-waste recycling, and maybe use them to power your next solar device. ◑

F CT scan of a larger capacity vape.

G

H

For a video demonstration and more projects from Becky Stern, visit beckystern.com/2023/05/28/reusing-disposable-vape-batteries

TIME REQUIRED: 1 Hour

DIFFICULTY: Easy

COST: $1–$5

MATERIALS
» **Acrylic sheet, 1/8" thick** in clear and color
» **Acrylic sheet, 3/16" thick** in colors
» **Adhesive label stock**

TOOLS
» **Acrylic solvent cement** with needle applicator
» **Computer and printer**
» **Laser cutter**
» **Hobby knife** e.g., X-Acto

Written, photographed, and illustrated by Bob Knetzger

Mystic Emoji Fortune Teller
Make this pocket-sized prognosticator and ask away!

BOB KNETZGER is a designer/inventor/ musician whose award-winning toys have been featured on *The Tonight Show*, *Nightline*, and *Good Morning America*. He is the author of *Make: Fun!*, available at makershed.com and fine bookstores.

A

B

Here's an update to the classic fortune-telling ball which you can easily make from acrylic.
Same fun play pattern: first ask a question, then shake and tilt to see your answer. But instead of text answers in a murky, fluid-filled ball, this flattened mini-version has emoji faces. For the top label design, choose a classic billiard ball or Makey's Crystal Ball — or create your own!

The design is a simple three-layer sandwich of acrylic (Figure **A**). The middle layer has a cutout

where the loose emoji disks slide around. When you shake and tilt the 'teller, one random disk slides down into the slot where it's revealed by the hole in the label.

MAKE IT
Go online at makezine.com/go/emoji-fortune-teller to download the .svg files for laser cutting. I used a Glowforge laser cutter which automatically sets the power and speed settings to cut the

different thicknesses of their Proofgrade acrylic. (The bottom and middle layers are made of thick ³⁄₁₆" acrylic. The clear top and the small disks are cut from thinner ⅛" acrylic. The Glowforge also automatically sets the depth of engraving of the emoji face designs.

Of course with a little trial and error any laser cutter will work. The design is simple enough to be scaled up and could be made by hand, but laser cutting quickly creates perfect miniature parts.

Finish the engraved emoji disks with some black acrylic paint (Figure **B**). Fill the engravings and then wipe off the surface for a clean graphic.

To assemble, bond the middle layer to the back with acrylic solvent cement (Figure **C**). Add the loose disks and then cement the clear cover on top (Figure **D**). Be careful not to get any solvent on the disks — they need to slide freely inside.

Print the label on some peel-and-stick label paper, then trim and cut out the viewing hole (Figure **E**). Align the hole with the slot and affix the label to the clear cover.

It's a perfect pocket size. Put it on a key ring or necklace. Or add a ribbon and hang it as an ornament on a holiday tree (Figure **F**).

SHAKE IT

Now for the fortune-telling fun. Ask a Yes/No question: Will you get your Christmas wish? Will your New Year's prediction come true? All will be revealed!

There's a 50/50 mix of Yes/No emoji outcomes (with two neutral):

If you want to temper the fates, you can change the mix of Yes/No emojis to create a more (or less) "opti-mystic" fortune teller.

Will you have fun making and playing with this mini-fortune teller? Signs point to Yes! ◐

C

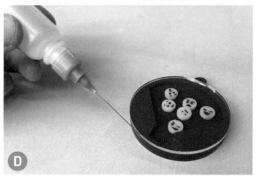

D

E

F

TILES AND TESSELLATIONS

Want to design interlocking geometric surfaces? A little trigonometry has you covered

Written by Joan Horvath and Rich Cameron

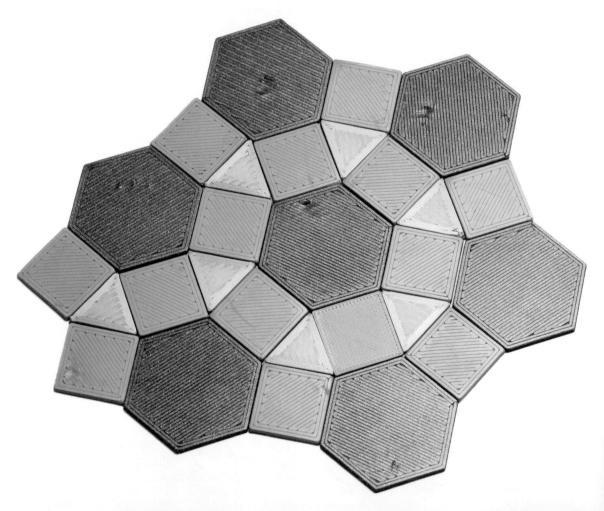

JOAN HORVATH and **RICH CAMERON** are the co-founders of Nonscriptum LLC (nonscriptum.com) and the authors of many books, including *Make: Geometry*, *Make: Calculus*, and the new *Make: Trigonometry*.

THE HIDDEN MATH OF BATHROOM FLOORS

Talk about tiling, and the first thing that comes to mind (for non-mathematicians, anyway) is an ordinary bathroom floor (Figure **A**). However, coming up with shapes that will cover a flat surface in an efficient and visually pleasing way is not as easy as it seems. Let's explore tiling with simple polygons and then jump to a newly discovered way to cover a surface called an "einstein monotile" (no relation to the physicist).

To a mathematician, ***tiling*** is the process of covering a surface completely with one or more repeating shapes, such that the shapes do not overlap or have any gaps between them. This is also called a ***tessellation.***

It might seem that it should be possible to tile a flat surface with any old regular polygon. (A ***regular polygon*** is a shape with straight sides, all of which are the same length and have equal angles between them, like a square or hexagon.) However, only triangles, squares, and hexagons can tile without other shapes taking up the spaces in-between. Pentagons and regular polygons with seven or more sides cannot tile by themselves. To understand why, let's remind ourselves about the features of a hexagon with six identical sides.

TILING WITH POLYGONS

A regular hexagon (Figure **B**) is a six-sided polygon. The triangle shown inside it is one of 12 identical triangles we can create inside the hexagon, two for each of the sides. Since all the triangles are identical, the ***interior angle*** of each will be 360° / 12, or 30°. We know that the other angle of the triangle is a right angle, and the angles of a triangle always add up to 180°. So that means that the angle that is half the ***vertex angle*** of the hexagon is 180° – 90° – 30° – 60°. The total angle inside each vertex, then, is twice that, or 120°.

If we look back at our floor tiles in Figure A, we see that each vertex is the intersection of three hexagonal tiles. We could draw a circle around each vertex, and see that the angles have to add up to 360°, a full circle (Figure **C**). Thus in the case of a hexagon, where each vertex angle is 120°, we can fit three of them inside this vertex

TIME REQUIRED: 1–2 Days + Print Time

DIFFICULTY: Easy/Moderate

COST: A Few Dollars

MATERIALS
» **3D printer filament** A small amount. This is a good project to use up some of those 50g samples!

TOOLS
» **3D printer** or online 3D printing service
» **Computer with OpenSCAD software** free download from openscad.org/downloads.html

Hexagonal tiles on a bathroom floor.

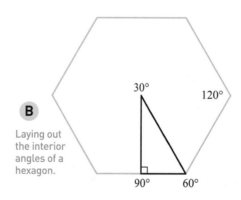

B

Laying out the interior angles of a hexagon.

30°   120°

90°   60°

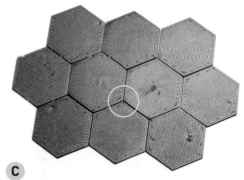

C

Hexagonal tiles with angles adding to 360° at a vertex.

D Squares adding up to 360° at a vertex.

E Triangular tiling.

F Tiling attempts with pentagons.

circle perfectly. Square tiles have vertex angles of 90°, so four of them come together at each vertex (Figure **D**). Six triangles come together at each vertex in a triangular tiling (Figure **E**).

IMPOSSIBLE TILINGS

But what happens when we get to five sides (a pentagon)? The interior angle is 360° / (2 * 5), or 36°. That means that half the vertex angle is 180° – 90° – 36° = 54°, and the vertex angle is twice that, or 108°. That means that 3.3333 pentagons would fit around a vertex — so it does not work. There is a good discussion of this problem in the 2017 *Quanta* magazine article by Natalie Wolchover referenced at the end of this article. Figure **F** shows one problematic tiling attempt.

The formula for the vertex angle in general, for a regular polygon of n sides, is:

$$vertex\ angle = 2(90° - (360°/2n))$$

As we saw, a hexagon does happen to work out. For regular polygons with more than six sides, though, a mix of other shapes (which might not be regular shapes, with all their sides the same) are needed to create a tiling. **Table 1** lists values of vertex angles for polygons with three to 10 sides.

TABLE 1. VERTEX ANGLES OF REGULAR POLYGONS

Polygon	Sides	Vertex angle (degrees)
Equilateral triangle	3	60
Square	4	90
Pentagon	5	108
Hexagon	6	120
Septagon	7	128.57
Octagon	8	135
Nonagon	9	140
Decagon	10	144

Now, if we no longer limit ourselves to tilings with all the same shape, and allow more than one, more possibilities open up. For example, suppose we used one square (90°) plus two octagons (135° * 2 = 270°). That adds up to 360°, and we can see in Figure **G** that a pattern of two octagons and one square (all with the same length sides) will meet at the vertices exactly. This is a common commercial tiling pattern too — watch for it now that you know about it.

Of course we can also come up with triangles or other shapes that no longer have all the sides of the same length to try and make tilings work.

PRINTING POLYGONS

It's easy to generate 3D-printed or laser-cut polygons with the free, open source CAD program OpenSCAD, downloadable from openscad.org. (Or you can make them in your favorite CAD program.) There is good documentation on that website, and we walk through using OpenSCAD in

G Octagons with squares.

detail in our *Make: Trigonometry* book. Briefly,

- Download and install OpenSCAD (for Mac OS, Windows, or Linux)
- Open a new file (Menu → File → New)
- Type in the short code that follows
- Render it (Design → Render)
- Export a 3D-printable STL file (File → Export → Export as STL).

If however you would prefer to just print the STLs, we have included STLs for a single triangle, square, pentagon, and octagon, as well as files for the two OpenSCAD models we describe here. You can download them all at makezine.com/go/skill-builder-tiling.

To print a regular polygon of thickness *t*, with *n* sides of length *s* millimeters, just type this code into OpenSCAD, or download our model *single_shape.scad*. (Set **t = 0** if you want to export an *.svg* file to laser cut or print in a 2D printer.)

```
s = 20; // length of a side, mm
n = 3; // number of sides
t = 2; // thickness in mm - 0 for 2D

if(t) cylinder(r = s / (2 * sin(180 /
n)), h = t, $fn = n);
else circle(s / (2 * sin(180 / n)), $fn
= n);
```

If you want to make multiple copies of one polygon in one output file, download the file *multiple_shapes.scad* instead. The parameters **x** and **y** are the maximum dimensions (in mm) we want to cover with our polygons for printing. To print them out on paper, U.S. copy paper is 8.5 by 11 inches, with a printable area of roughly 200 by 260mm; as before, set **t = 0**. For 3D printing, set **x** and **y** to the dimensions of the printable area of your print bed (or smaller). This model will lay out a reasonably efficient grid of shapes, with a bit of margin, into the specified space.

```
s = 20; // length of a side, mm
n = 4; // number of sides
t = 2; // thickness in mm - 0 for 2D
// Paper or print bed dimensions
x = 200;
y = 200;
```

H Pentagon and decagon tiling.

I Triangle, square and hexagon tiling.

```
for(
  x = [0:s / sin(180 / n) + 1:x - s /
sin(180 / n)],
  y = [0:s / sin(180 / n) + 1:y - s /
sin(180 / n)]
)
  translate([x, y]) rotate(180 / n)
  if(t) cylinder(r = s / (2 * sin(180 /
n)), h = t, $fn = n);
  else circle(s / (2 * sin(180 / n)),
$fn = n);
```

Print yourself a set of various polygons and try out tilings. The vertex angles in Table 1 might help with your brainstorming. For example, we might expect two pentagons and a decagon meeting at each vertex to work, based on the sum of the angles. However, if we try it (Figure **H**) it turns out that we wind up with a ring of pentagons around each decagon, with gaps between the pentagons. It is necessary for the vertex angles of tiles coming together to create a full circle, but that doesn't guarantee that those shapes can tile a whole surface without gaps. Other tilings are possible by mixing triangles, squares, and hexagons (Figure **I**).

APERIODIC TILINGS

Not every tiling is a pattern that repeats. If a tiling is a result of a set of shapes that repeat over and over but do not form a repeating pattern, this is called an *aperiodic tiling.* There are a number of these, and it has been a bit of a mathematical competition for centuries to come up with interesting tilings that use very few pieces over and over. Many of these tiles have concave vertices (cupped inward from the polygon's surface) in addition to the convex vertices (bulging outward) we are used to seeing in our regular polygons.

For example, one type of *Penrose tiling* — named for Nobel-winning physicist Sir Robert Penrose who investigated them in the 1970s — uses just two shapes, called the dart and the kite, to cover a surface aperiodically.

For some time, the holy grail of aperiodic tiling was to find an *einstein tile* (from the German "one stone," not the famous physicist). An einstein, or *aperiodic monotile,* would tile a whole surface using a single tile shape, in a way that did not repeat. In 2023 the team of David Smith, Joseph Samuel Myers, Craig S. Kaplan, and Chaim Goodman-Strauss discovered an einstein that can cover a surface without ever repeating a pattern (albeit needing to be flipped over in some places). They dubbed it the "hat" (Figure **J**) although some people think it looks more like a T-shirt. Subsequently, they have come up with a different version that appears to work without ever flipping it over!

The hat is an irregular shape with 13 sides. Three of them put together are shown in Figure **K**, and a larger set in Figure **L** where the blue tiles have been mirrored (flipped over).

You can read about the hat and the effort to prove that it does what the authors think it does at cs.uwaterloo.ca/~csk/hat, which has links to a preprint paper and other resources. English mathematician Christian Lawson-Perfect has created a repository of various ways to print out the hat and other tiles, in OpenSCAD or a variety of 2D and 3D printable formats, which you can find at github.com/christianp/aperiodic-monotile. The models in Figures J, K, and L are printed from his OpenSCAD file.

Building out a pattern is harder than it looks,

J The "hat" einstein tile.

K Three hats fitted together.

L Pattern formed with many hats, some of them flipped.

and pretty addictive. Try to look at the vertices and reason about what is happening at each one. A convex vertex should be thought of as an angle greater than 180°. For the vertices where a tab pokes into a concave area on the other tile, the angle of the convex tab plus that of the hole around it will add up to 360°. There are

Rich Cameron

other variants to explore discussed at the links, including one that does not have to be flipped over to tile the whole surface. Mathematicians are reviewing the work, and there should be a lot more written about it soon.

MORE TILING PUZZLES

Finally, although it is not strictly speaking a tiling, you might want to play with a 2,000-year-old tile puzzle attributed to our friend Archimedes, the **ostomachion** (see en.wikipedia.org/wiki/ostomachion). You can cut out its 14 pieces from one of the versions online and explore the various ways the triangles and quadrilaterals can be put together. People have also created versions of it on the model-download sites printables.com (as "Archimedes puzzle") and thingiverse.com.

Similar 7-piece puzzles called **tangrams** have been around in China for millennia, and have been popular in the West as entertainments for a long time, too. They (and the ostomachion) belong to a broader class called **dissection puzzles**, if you want to explore even more. Playing with one (or, even better, trying to create one) is a great way to build your intuition about relationships among different simple shapes.

TILING IN THE WILD

Similar repeated patterns arise in many other contexts beyond the bathroom floor. For example, close to home for 3D printing folk, hexagonal infill is a strong and stable way to create the structure of a 3D print. Figure 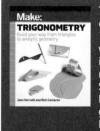 shows a square part printed with zero top and bottom layers, and 20% hexagonal infill. And of course a beehive is a closely-packed set of hexagons.

Tessellated patterns have appeared in art, architecture, and engineering since ancient times, in two dimensions and in three. In three dimensions, tiling on a variable surface is a bit of a different situation than tiling a flat floor, but the general premise is the same. If you have used an STL file for 3D printing or computer aided design (CAD), the acronym stands for either *stereolithography* (a 3D printing technology) or *standard tessellation language,* depending on who you ask. Tessellation is another word for tiling. STL files consist of many triangles and their orientations. When you get a complaint from

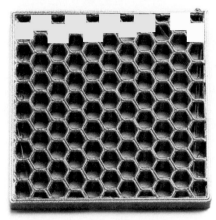

M Hexagonal infill in a 3D-printed part.

your slicing program that an STL has problems, it is probably because there are holes where the triangles are not completely tiling the surface.

Moving on from tiling per se, if you find this process of packing simple shapes into complex repeating patterns intriguing, you might also want to explore the different types of crystals, which pack molecules into 3D shapes. As we explored in our 2016 Apress book, *3D Printed Science Projects,* water ice crystals can be hexagonal in shape. Look up "crystal structures" to learn about these different types of repeating three-dimensional lattices.

LEARNING MORE

A good general introduction to tiling can be found in Natalie Wolchover's article in *Quanta* magazine (July 11, 2017), "Pentagon Tiling Proof Solves Century-Old Math Problem," quantamagazine.org/pentagon-tiling-proof-solves-century-old-math-problem-20170711. ◗

This article is adapted from our upcoming book, *Make: Trigonometry*. Trigonometry was the original maker math, immediately useful in navigation and surveying. See how to turn trig into a hands-on experience by grabbing a copy from the Maker Shed at makershed.com/products/make-trigonometry-print.

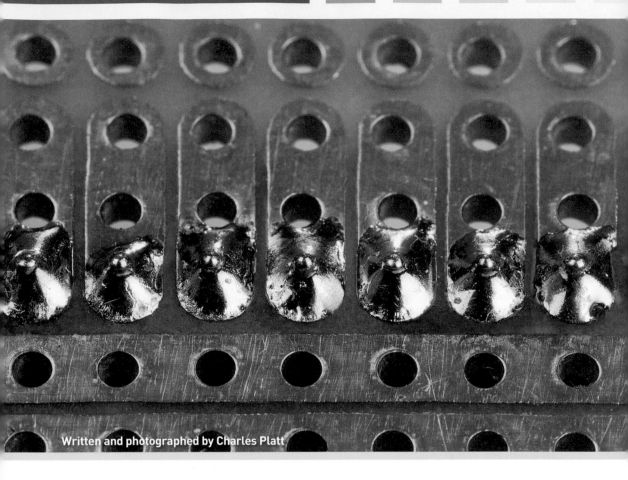

Written and photographed by Charles Platt

SERIOUS
SOLDERING

Banish bad habits and solder like a boss with
an inexpensive temperature-controlled iron

CHARLES PLATT is the author of the bestselling *Make: Electronics*, its sequel *Make: More Electronics*, the *Encyclopedia of Electronic Components Volumes 1–3*, *Make: Tools*, and *Make: Easy Electronics*. makershed.com/platt

For many years, I was in denial about my soldering. I knew how it should be done, and even wrote about it in my electronics books; but I didn't enjoy doing it, and my impatient attitude created results that were — well, a bit embarrassing.

I told myself that it didn't matter. So long as my solder joints conducted electricity and didn't fall apart, why should I care if they were large and lumpy?

Secretly, of course, I did care, and my denial was always a fragile facade. It was finally demolished when I took on a project requiring a lot of very precise soldering, and I couldn't pretend anymore. I had to get serious.

GOING THERMOSTATIC

First I addressed the issue of temperature. I always used to reject **thermostatic soldering** as an unnecessary expense, but with irons such as the one in Figure Ⓐ selling for as little as $15, there was no excuse for not buying one.

Now that I could set the temperature, what should it be? I found a huge amount of unsubstantiated internet folklore on this subject, so I checked with manufacturers of hand-soldering production-line equipment to get authoritative advice. They recommended **325 to 350 degrees Celsius** (617°–665°F).

That seemed excessive, as solder containing 60% tin and 40% lead melts at only 185°C. Lead-free solder requires about 220°C, but still, that is far below 325.

The reason for using a higher temperature is that it helps you to work fast, completing each joint in 3 seconds or less. A lower-temperature soldering iron will take longer to make the joint, because heat flows into it more slowly. But this allows extra time for heat to spread into locations where you don't want it, such as sensitive semiconductors. Prolonged heat will also melt the insulation off wires.

I chose a soldering temperature of 340°C. Now, was the iron accurate? I attached a thermocouple to the tip, using a tiny loop of fine steel wire, and Figure Ⓑ shows some results from a data logger. My low-priced thermostatic iron approached its target temperature in only 90 seconds, while an old generic soldering iron that I had owned for many

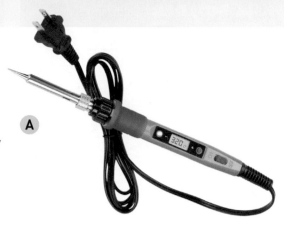

Ⓐ

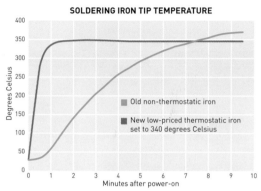

Ⓑ Tip temperatures of soldering irons. 21 data points, with some smoothing.

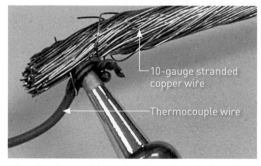

Ⓒ Abusing a soldering iron by draining its heat into 10-gauge copper.

years needed more than 7 minutes to catch up — and then became hotter than I wanted it to be.

My testing setup was imperfect, because some heat was lost into the thermocouple wiring. But thermostatic control compensates by increasing power to the heating element in response to demand. In fact, my cheap little iron claimed to have a 90W power supply, so I decided to test its limits.

I held a piece of 10-gauge stranded copper wire against the tip, as in Figure Ⓒ. The blue line

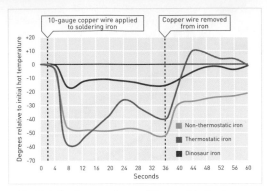

D Relative temperatures of three soldering irons subjected to heat abuse. 16 data points in each curve, with some smoothing.

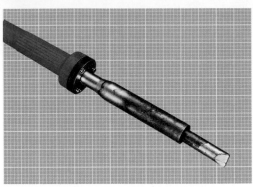

E This thing is big. The background grid is in inches.

F Homemade helping hands. Total materials cost: Under $2.

G Old-style helping hands, secured with wing nuts.

H These Quad Hands are expensive, but a pleasure to use.

in Figure **D** shows the result: The wire drained so much heat, the temperature plummeted by 60°. The switching power supply inside the iron eventually restored the correct temperature, but thermostatic irons may be as guilty as cheap stereo systems when it comes to making exaggerated claims for wattage.

To solder heavy wiring, such as you might use for amp-hungry 12VDC motors in a large robot, I still prefer an old "dinosaur iron" such as the one shown in Figure **E**. Its combination of heat and weight can be described as *thermal mass*, and the red curve in Figure D shows how well this thermal mass is able to withstand sudden demands.

For detail work, such as soldering a circuit board, I have to conclude that modern thermostatic irons are a good low-cost choice.

GETTING A GRIP

Because my state of denial about soldering always discouraged me from spending money on it, I had improvised my own *helping hands* from Romex house wiring epoxy-glued into a chunk of 2×6 lumber, as shown in Figure **F**. Two alligator clips could hold a circuit board while the third gripped each component that I was attaching to it.

This was slightly more convenient than the old, basic helping hands, with their nasty little wing nuts that always seem to loosen themselves (see Figure **G**). But now that I was cleaning up my soldering act, I wanted something better.

My final choice was the set of Quad Hands shown in Figure **H**. Although they are five times as expensive as the basic ones, they are ten times as convenient and versatile. I started to enjoy using this well-made device, after I got over paying for it.

FEELING THE FLUX

I always thought a separate supply of flux was just for plumbers, since electronic solder has its own flux in a hollow core. What I failed to realize (until a friend made it clear to me) is that if you take too long to make a joint, the flux in solder can boil off, with nasty consequences.

See for yourself. Try melting a quarter-inch blob of flux-cored solder on the spade-shaped tip of a medium-power iron, and let it sit there. The fumes you see are the flux going up in smoke. Wait 30 seconds for the fumes to stop completely, and then apply the solder to a thick piece of wire. Heat it for as long as you like; the solder will end up looking like the disaster on the right-hand side of Figure **I**.

This is why solder should be applied to the joint, not to the iron, and as quickly as possible. After solder loses its flux, it will not bond properly with copper. You'll be able to scrape it off afterward with your fingernail, especially if it's lead-free solder.

Flux is often described as a ***wetting agent***, meaning that it reduces the surface tension of the liquid solder, so that it spreads as a film. Flux also cleans oxide deposits off copper, enabling a chemical bond to occur.

Rosin flux for electronics is sold in little tubs, or as liquid in a syringe, or in pens for painting it onto circuit boards, as in Figure **J**. ***No-clean flux*** can remain as a residue on a circuit board, while other types should ideally be removed. Kester, a flux manufacturer, is a great source of information: kester.com/knowledge-base.

A HOT TIP

After I corrected most of my bad habits, one still remained: I was lazy about ***cleaning the tip*** of my soldering iron. Maintaining a damp sponge seemed too much of a hassle—until I read a dissertation by an engineer at a soldering iron manufacturer, who said you should sponge-wipe an iron after every single joint.

Seriously? Yes, like most engineers, he was very serious. I never liked dinky little sponges, so I wadded up a pure cotton rag, put it in a ceramic bowl, and moistened it with distilled water. That has been a big success, so long as I remember to use it. Now each of my soldering irons has a

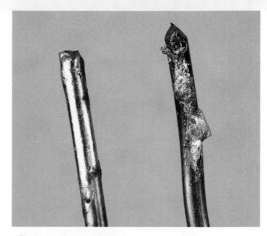

I Left: Solder applied with flux. Right: Without flux.

J Fluxes in a little tub, in a syringe, and in a pen.

gleaming silvery tip, and — yes, they transfer heat much more effectively than when they were dirty.

BANISHING THE BAD HABITS

I'm happy to report that my bad habits are history. Since my recovery from bad-soldering denial, I actually enjoy soldering, as a craft in itself. I don't get impatient, and I don't have to hide my solder joints in shame.

I think it's like cooking. You can roast some beef in your oven till it looks about right, and hope for the best — but if you want predictable, repeatable results, you'll use a timer and an oven thermometer.

I always knew this.

I just had to get serious enough to stop pretending. ◉

• • • ***With thanks to Ken Coffman, flux guru.*** • • •

Bambu X1–Carbon With AMS

$1,449 (as built)   us.store.bambulab.com/products/x1-carbon-combo

The X1-Carbon made ripples in the 3D printing community when Bambu seemingly came out of nowhere and launched a printer that boasted speeds that felt unreal — three times faster than typical FFF machines — with a plug-and-play system to print in multiple colors, and claimed it was all very user friendly. We bought one to test it out and see if the hype is justified.

I've been using this machine non-stop for weeks, and the experience is fantastic.

Using both PLA and ABS, my prints look fantastic and finish super fast. The system has a tiny lidar scanner that inspects the entire first layer to detect any issues. It works great, though I do find I need to wash my plate and apply glue stick fairly regularly to keep that first layer going down flawlessly.

The Automatic Material System is where the Bambu really shines and is by far the best multi-

material experience I've had. Selecting colors in the slicer is simple thanks to Prusa's open-source slicer development. You can literally just paint over areas you want to be different colors, or select polygons individually. You can have four different filaments loaded and waiting, or even chain together multiple AMS systems for up to 16 colors.

There are a few downsides to the AMS. There's a ton of waste — each color change produces a purge blob called a "poop" and takes a bit of time. This adds up quickly.

Ultimately, I'm super impressed by this machine. It has earned a spot next to my fleet of Prusa printers for the highly demanding parts I print for my gaming charity. That's rare — no other FFF machine has performed well enough to sit next to my MK3 for constant precise parts, all day every day. —*Caleb Kraft*

Sumit Basra, Caleb Kraft

Wainlux K8

$419–$559 kickstarter.com/projects/
wainlux-laser-tech/wainlux-k8-the-fast-
and-powerful-laser-engraver

The Wainlux K8 is an interesting approach to the desktop hobby laser market. At 10W, fully enclosed, and retailing around $500, it's firmly aimed at buyers without tons of space or a big budget.

I hooked it up to LightBurn software and without any real tweaks I was engraving and cutting without issue. The included CutLabX software is a bare minimum; I'm not even sure it has an Undo feature. I also couldn't get the K8's camera to work from CutLabX or LightBurn; hopefully Wainlux resolves these issues.

I like how they handle focusing the laser beam — a must for a perfect cut — using an easy-to-move bed and a flip-down distance gauge attached to the laser head.

The air filtration system worked surprisingly well. It's fairly easy to clean out and the built-in air assist gives you a cleaner cut and less debris buildup, without having to attach an external pump or hoses. You'll still smell burnt material as there are gaps in the enclosure (but I had less than from my K40, even with its bigger exhaust).

I really like this approach. While the small operating area is a bit limiting, the fact that it is fully enclosed and fairly portable means I can store it away between uses and even travel with it. The 10W laser is powerful enough for most tasks I need at this scale. —CK

Phrozen Sonic Mini 8KS

$350 phrozen3d.com/pages/sonic-mini-8k-s

Phrozen's latest resin printer boasts super fine print quality at a very affordable price. With 22μm resolution on X and Y axes, you'll see incredible detail and features on your prints. It's difficult to find any voxelization even when you're looking for it!

I loaded up a digital sculpt I made of a realistic Gamera-style turtle monster, and the final result is stunning. I usually worry about fine detail but looking at this, I'd say I could even add more detail! However, the prints stick to the build plate almost too well. You really need to fight to get them off, chipping away the raft slowly.

The included slicer, a trial version of VoxelDance Tango, was easy enough to use but I'd prefer a full slicer, even if it were less functional. You can also use free slicers like PrusaSlicer with this machine.

Despite a few very minor annoyances (the loud beep when starting a print, and that sticky build plate texture), the price makes this a no-brainer. The print quality coming off this machine is just crazy good. —CK

Beagle Camera V2

$95 mintion.net/products/mintion-beagle-v2-3d-printer-camera

Mintion's upgraded remote 3D printer monitor has much more capable focus and much higher video quality. It offers the ability to control your printer remotely, queue up jobs, see what's happening in real-time, and record time-lapse videos.

Setup is easy and being able to keep an eye on your prints is very convenient. The timelapse feature is OK; the default method isn't the best for a bed slinger like the Prusa, but with a bit of tinkering you'll get there. And it's still useful for troubleshooting failed prints.

Overall, the interface is simple and results are nice. If remote view and control are worth $95 to you, the BeagleCam V2 delivers easily and quickly. —CK

The String and Glue of Our World: Understanding Composite Materials

$30 nedpatton.com

Fiberglass boats, carbon rods, stealth airplane skins, and plain old plywood — they're all composite materials: some kind of fiber reinforcing a matrix of some kind of resin. Mechanical engineer Ned Patton spent 40 years in the field, then wrote this eye-opening and highly readable guide to composites. It's a unique book. Yes, we get a deep dive into how composites work, at the molecular level (hint: covalent bonds) and also at the design and production levels (CAD tools, fiber weaves, lamination, molding, vacuum bagging), as well as how they don't work (failure modes). But along the way Patton also provides an enjoyable survey and history of these materials — from familiar resins like polyester and epoxy to exotic fibers like Kevlar, Nomex, Dyneema, and Vectran — and then caps it off with predictions of where the industry is going next and how you can get in, whether it's higher education or straight to the job market.

Recommended for anyone who wants to get their head around the composite materials of today and the future.

—Keith Hammond

ROBOT KITS

We do love robot kits! Here are a few that hit the beginner-to-intermediate sweet spot, to get you started and then help you advance your robotics skills.

COMPLETE KITS

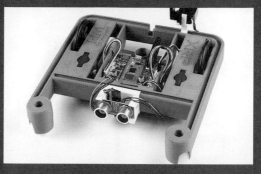

Makey:Bit–Mobile Kit

$125, discounts available
makershed.com/products/makey-bitmobile-kit

Make:'s own kit is a two-fer, giving you a pair of BBC Micro:bit computers and a pair of Makey:bit Adventure Boards — our carrier board for Micro:bit, BrainPad, or Adafruit Clue, each with twin servo connectors, I²C port, full pin access, battery holder, and onboard LEDs. Plus all the parts you need to build a two-wheeled robot and remote controller: two servos, 3D-printed wheels with TPU treads, marble caster, joystick, pushbutton, laser-cut frame parts, and all hardware and cables. It's a good buy and a great entry point for learning coding, physical computing, and robotics using the kid-friendly block-based MakeCode platform. —*KH*

XRP Experiential Robotics Platform

$115 experientialrobotics.org

This open source kit from SparkFun is tailor-made to get students into robotics using the same lessons and platforms as spendy FIRST Robotics programs, but for a fraction of the price.

Built around a Raspberry Pi Pico W microcontroller, the XRP Controller Board has Wi-Fi and Bluetooth for connectivity, a high-quality IMU chip for motion measurement, four motor controller ports, two servo ports, and a Qwiic connector for sensors, LCDs, and such. Beginners can code it in Blockly, or move up to languages like MicroPython and Arduino. The kit ships with 3D-printed robot parts (also shared on GitHub) plus an ultrasonic rangefinder, line follower, two motors with encoders, servomotor, Qwiic cables, wheels, casters, and battery holder.

With a curriculum from Worcester Polytechnic Institute (WPI) and a list of partners including DigiKey, Raspberry Pi, MatterHackers, and DEKA, the XRP kit (SparkFun 22230, DigiKey KIT-22296) is meant to make an impact in education and serve as a bridge to FIRST Robotics. —*KH*

mBot Neo

$130 makeblock.com/pages/mbot-neo-coding-robot

Makeblock's mBot Neo is an adorable, user-friendly robot that makes learning fun with its cutting-edge technology, real-world applications, and Lego compatibility. Using the included tools and clear instructions, my daughter easily assembled the robot in less than a half hour. It's sturdy and compact, with an aluminum frame and durable, high-quality plastic parts. The ESP32-based CyberPi main board doubles as a removable remote controller, a clever way to keep the excitement going — once kids have completed the build, they can take a play break, drive the mBot around, and get excited about coding it. The unit features a color display, gyro/accelerometer, ultrasonic and light sensors, expansion ports, rechargeable battery, and onboard LEDs, microphone, Wi-Fi, and Bluetooth. It even supports voice commands and AI/machine learning.

mBot's Scratch-based coding platform mBlock allows kids to simply click to assemble blocks of code without fear of typing errors. As they progress, they can tackle "real" languages like Python or Arduino C. The mBot's ease of use and range of applications has made it a regular guest on playdates at our house with kids from ages 5 to 16. Save $10 with promo code *Make10*. —*Gillian Mutti*

CARRIER BOARDS

Adafruit CRICKIT

$30 adafruit.com/category/996

This do-it-all robotics carrier board is available for Feather, Circuit Playground Express, Micro:bit, or Raspberry Pi. It's packed with ports — four servos, two DC motors at 1 amp each, four high-current outputs for solenoids and such, capacitive touch sensors, digital I/O, NeoPixel LED, and even 3W audio. And you can control them all with just two data pins via Adafruit's "Seesaw" I²C chip! With built-in overcurrent, overvoltage, and motor kickback protection, plus tons of tutorials, CRICKIT is a well-thought-out and durable board to roll your own robots. —*KH*

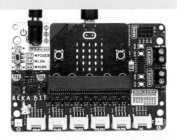

Reka:Bit

$18 + shipping from Malaysia cytron.io/p-rekabit-simplifying-robotics-w-microbit

A capable carrier for Micro:bit, with ports for four servos and two DC motors, onboard LEDs, and six Grove ports for I²C, analog, and digital I/O. It's well reviewed and a good value if you're buying a batch. —*KH*

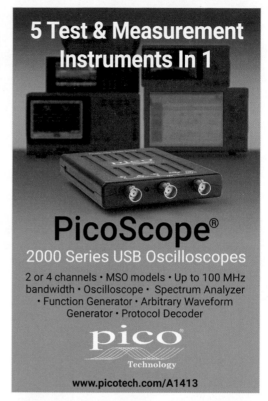

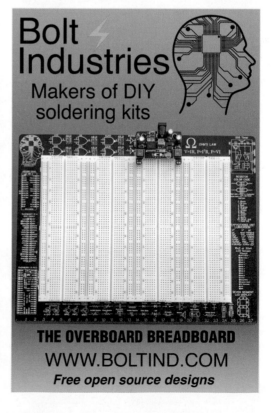

Courtesy of Empire of Dirt

TO THE STARS AND BACK

On August 26, 1965 NASA launched Project EMPIRE, a secret, manned mission to Mars, but two months into their journey all communication with the spacecraft ceased and the mission was considered lost. Or were they?

Masterminded by Jon Sarriugarte and Kyrsten Mate of the art group Empire of Dirt, *Project EMPIRE* represents an imagined far future restoration of an "ancient" 1960s spacecraft across different timelines, with the original crew working with their distant descendants to bring the ship back home to Earth. Without replacement parts, they've had to improvise and adapt new materials, like an alien bug carapace (which is really made from welded steel and fiberglass), to make the ship space-worthy.

For the more pedestrian-minded, the 18-foot-tall, 35-foot-long art car is built on top of an Isuzu NQR diesel-powered truck. A total of 25 artists, fabricators, and students worked on the "aerospacepunk" project off and on for the past 8 years. But it's the attention to detail the team is really proud of: from the hotrod-style paint job and meticulously detailed engine to the intricate sound design and deep backstory at projectempire.org, everything has been crafted to look as real and functional as possible.

—Craig Couden